Biological and Medical Physics, Biomedical Engineering

More information about this series at http://www.springer.com/series/3740

BIOLOGICAL AND MEDICAL PHYSICS, BIOMEDICAL ENGINEERING

The fields of biological and medical physics and biomedical engineering are broad, multidisciplinary and dynamic. They lie at the crossroads of frontier research in physics, biology, chemistry, and medicine. The Biological and Medical Physics, Biomedical Engineering Series is intended to be comprehensive, covering a broad range of topics important to the study of the physical, chemical and biological sciences. Its goal is to provide scientists and engineers with textbooks, monographs, and reference works to address the growing need for information.

Books in the series emphasize established and emergent areas of science including molecular, membrane, and mathematical biophysics; photosynthetic energy harvesting and conversion; information processing; physical principles of genetics; sensory communications; automata networks, neural networks, and cellular automata. Equally important will be coverage of applied aspects of biological and medical physics and biomedical engineering such as molecular electronic components and devices, biosensors, medicine, imaging, physical principles of renewable energy production, advanced prostheses, and environmental control and engineering.

Kirill Kulikov · Tatiana Koshlan

Laser Interaction with Heterogeneous Biological Tissue

Mathematical Modeling

Second Edition

 Springer

Kirill Kulikov
Peter the Great St. Petersburg
  Polytechnical University
St. Petersburg, Russia

Tatiana Koshlan
Saint Petersburg State University
St. Petersburg, Russia

ISSN 1618-7210   ISSN 2197-5647   (electronic)
Biological and Medical Physics, Biomedical Engineering
ISBN 978-3-030-06799-1   ISBN 978-3-319-94114-1   (eBook)
https://doi.org/10.1007/978-3-319-94114-1

This Springer imprint is published by the registered company Springer International Publishing AG
part of Springer Nature
The registered company address is: Gewerbestrasse 11, 6330 Cham, Switzerland

Preface

One of the most important areas of application of laser radiation is biomedical optics. Here, laser sources are used for diagnosis, therapy, or surgery.

Note that for the development of new methods of laser biomedical diagnostics, a detailed study of the propagation of light in biological tissues is required, as theoretical studies improve the understanding of the optical measurements, increase capacity, reliability, and usefulness of optical technologies.

To solve these problems in the first place, the most informative indicators of the functioning of the organism must be chosen. These indicators are the results of the analysis of peripheral blood. Peripheral blood provides the complete information on the status of the human organism. A comprehensive study of the characteristics of light scattering and absorption can quickly detect intact physiological and morphological changes in the cells due to thermal, chemical, antibiotic treatments, etc.

The choice of the laser beam to study the structure of biological particles is conditioned by the fact that it does not induce gross pathological changes, and diagnostics will ensure effective use of all the coherence properties of laser radiation, monochromatic directional.

Note that for laser processing of the biological environment, it is also necessary to perform a selective thermal influence facility located in the environment. For these purposes, it is necessary for a selection of optimal spectral, temporal, and energy characteristics of the laser source.

The main parameter to reach selectivity (choice) is the wavelength of the radiation. If we choose a wavelength of light which is absorbed by the object, but not absorbed by the surrounding tissues, the selectivity is achieved.

However, such a situation is ideal and cannot always be achievable in practice. Considerable value is also placed on the duration of treatment, the size of the object, the depth of its location.

After laser irradiation on biological tissues, factors such as the movement of blood through the vessels and thermoregulation are need to be considered.

Blood flow can have a significant impact on the result of exposure if it is dependent on the degree of thermal damage to the tissue, because blood flow may be an additional and sufficiently effective mechanism for heat removal from the site of exposure. Note that this effect may influence both the efficiency and the safety of the procedure, because it violates the local heating.

Thus, the optimization of the laser emitter for selective heating of multicomponent media is an ambiguous problem.

For these purposes various mathematical models have been developed, which are usually designed to solve a specific task. In most cases, the problem of choice of the laser source and its performance are decided on the basis of the absorption and relaxation times of the objects (media). Modeling of this kind is usually designed to solve the problem of optimizing the parameters of the laser transmitter and evaluate the results obtained under the influence of the preselected laser on the biological environment. In order to correctly construct a mathematical model that describes the interaction of laser radiation with tissue, it is first and foremost a establish good understanding of the structure of biological tissues, their optical and thermal properties, as well as the main effects in the propagation of radiation in biological tissues.

The monograph discusses problems related to the study of mechanisms of interaction of laser radiation with biological tissues, the study of effects of laser interaction with biological tissues methods of the asymptotic theory of diffraction, and computer modeling. By virtue of models described in the monograph, on the basis of result of influence of laser biological tissue under certain conditions, can be consistently changed to input characteristics to produce an optimization of the spectral and energy parameters of laser emitters to achieve the desired effect in each case.

The book presents the original results of theoretical studies of electromagnetic waves in media-simulating biological-layered structure. Concepts and methods for studying the laser radiation interaction with multicomponent heterogeneous tissue with a complex structure of the asymptotic theory of diffraction methods are presented. These methods can serve as the basis for creating a software for the biomedical diagnostics.

The monograph is addressed to researchers and specialists in biomedical physics interested in the development and application of laser and optical diagnostic methods in medical research.

The monograph consists of ten chapters.

In Chap. 1, we consider the structure and optical properties of biological tissues, blood, and human skin.

In Chap. 2, we expand methods of light scattering for the quantitative study of the optical characteristics of the tissue, and the results of theoretical and experimental studies of photon transport in biological tissues.

In Chap. 3, we describe the optical characteristics, namely, dispersion and absorption spectra of an ensemble of spherical particles randomly oriented inside an optical cavity. The study is based on the self-consistent matching of new data from the inhomogeneous optical cavity with data from the scattering of an ensemble of spherical particles of different size, randomly oriented in free space.

In Chap. 4, we discuss a mathematical model for calculating the interaction of laser radiation with a turbid medium and a model for the prediction of the optical characteristics of blood (refractive index and absorption coefficient) and for the determination of the rate of blood flow in the capillary bed under irradiation of a laser beam is proposed.

In Chap. 5, we construct an electrodynamic model, which makes it possible to vary the electrophysical parameters of a biological structure in calculations with allowance for roughness.

In Chap. 6, the mathematical model is proposed for predicting optical characteristics (refractive index and absorption coefficient) of a biotissue being simulated, which is probed in vivo by a laser beam. Blood corpuscles, in this case, are simulated by particles of irregular shape and various sizes, which are oriented arbitrarily in free space.

In Chap. 7, the mathematical model is proposed for detection of the function of size distribution of form for blood cells. Using the mathematical model, we can theoretically calculate the size distribution function for particles of irregular shape with a variety of forms and structures of inclusions that simulate blood cells in the case of in vivo and determine the degree of aggregation, for example, the platelet for the in vivo case.

In Chap. 8, we construct a mathematical model, which allows us to vary the electrical parameters and structure of the simulated biological tissue with fibrillar structure for in vivo case.

In Chap. 9, we expand a mathematic model for predicting the absorption spectrum and dispersion of a section of a biological structure consisting of epidermis, upper layer of the derma, blood, and lower layer of the derma and placed in the cavity of an optical resonator.

In Chap. 10, we discuss a mathematical model, which makes it possible to vary the characteristic sizes of roughness, the electrophysical parameters of the biological sample under investigation, and its geometrical characteristics and to establish the relations between these parameters and biological properties of the biological tissue being modeled, as well as to calculate theoretically the absorption spectra of optically thin biological samples placed into the cavity of an optical resonator.

In Chap. 11, we propose a mathematical model for calculation of the hyperthymia of a multilayer biological structure under the action of laser radiation.

In Chap. 12, the mathematical model is proposed for determination of the optical parameters on the basis of spectrophotometric data and we consider general structure models' interaction of laser radiation with a biotissue.

Before closing, we want to acknowledge our sincere thanks to colleague Prof. A. P. Golovitskii for a critical reading of the manuscript. Our thanks are to Springer Nature, in particular, Dr. Habil Claus E. Ascheron.

St. Petersburg, Russia   Kirill Kulikov
                                                                    Tatiana Koshlan

Contents

Chapter 1
Methods Describing the Interaction of Laser Radiation with Biological Tissues

Abstract We consider the structure and optical properties of biological tissues, blood and thermophysical characteristics of the elements in the skin tissue.

1.1 Introduction

Scattering and absorption of electromagnetic radiation are widely used in various fields of science and technology to study the structure and properties of heterogeneous environment. Theoretical models, techniques of experimental research and methods of data interpretation were developed by specialists of various disciplines (from astrophysics to laser ophthalmology), so there are differences in traditions and terminology barriers that impede the efficient interaction of different schools of thought. For example, experts in the field of atmospheric optics and astrophysics use the ideology of natural radiation transport equation, but for interpretation of data, small-angle X-ray and neutron scattering - more familiar language - using the apparatus of the correlation functions and structure factor scattering.

Due to the large variety and structural complexity of the biological environment the development of adequate models of optical scattering and absorption of light is often the most difficult part of the study. These models cover almost all the major sections of optical dispersion media: a simple single-scattering approximation, incoherent multiple scattering, which is described by the transport equation, and the multiple scattering of electromagnetic waves in condensed systems interacting lenses or irregularities.

Before proceeding to the consideration of principles of mathematical models to calculate the interaction of laser radiation with biological structures and objects of different degrees of complexity and organization, consider the structure and optical properties of biological tissues.

© Springer International Publishing AG, part of Springer Nature 2018
K. Kulikov and T. Koshlan, *Laser Interaction with Heterogeneous
Biological Tissue*, Biological and Medical Physics, Biomedical Engineering,
https://doi.org/10.1007/978-3-319-94114-1_1

1.2 The Structure and Optical Properties of Biological Tissues

From the point of view of controlling the optical parameters of tissues the fibrous tissues (sclera eyes, dermis of skin, dura, etc.) are of the most interest makes. Fibrous tissue make up approximately 50% of body weight. Loose connective tissue of fatty tissue, dens, tendons and intermuscular fascial layers, derma of skin and an intraorganic stroma of parenchymatous organs, neuroglia and peritoneum are all connecting fabrics or fibrotic thanks to the existing characteristics of fibrillar structures [1]. All varieties of fibrous tissue in spite of their morphological differences, are built an common, with the same principles, which mainly include the following [1]: (a) the connective tissue contains cells, but compared to other tissues there are fewer. As a result, the amount of intercellular substance is larger than the cellular elements; (b) the connective tissue is characterized by the presence of fibrous (fibrous) structures - collagen, elastin and reticular fibers, which are the main structural elements of the fibrous tissue, and surrounded by intercellular substance; (c) the connective tissue is rich with the intercellular substance when has a very difficult chemical composition. The characteristic component of the structure in fibrous tissue (tendons, cartilage, the dermis of the skin, eye sclera, etc.) are the collagen fibers. They carry protective functions. For collagen, the specific amino-acid structure and a unique spatial location of polypeptide chains is a characteristic. As opposed to other proteins, large number of amino acids are contained in collagen: glycine, proline, hydroxyproline, lysine, oksilizin. Collagen makes up 25–30% of the total protein in adult, or 6% of total body weight [2]. The molecule of collagen consists of three polypeptide chains forming structure of a threefold spiral. The length of the collagen molecule is 280 nm, the diameter of 1.4–1.5 nm [3].

The Structure and Optical Properties of the Skin

Human skin is an example of a multicomponent turbid biological medium and it is very difficult to describe the construction of the models. Optical characteristics of such complex environment as a whole depend on many factors. To correctly build the model of the skin and the description of optical properties one needs to get some understanding in the biological features of the structure of skin (see Fig. 1.1).

Skin consists of two main layers [4]. The external layer is the multilayered flat keratinized epithelium - epidermis. The average thickness of the epidermis, of which there is relatively little change in thickness, is approximately 100 μm [5]. Epidermis consists of two various layers of cells: an external, corneal layer of dry acaryocytes *(Stratum corneum)* and inside cellular layers, i.e. actual epidermis from which, after modification, there are superficial cells: *(Living epidermis)* [6]. The main type of cells in this epitelialny continuum is the epidermalny cell, most often called keratinotsity, so called because the family of fibrous proteins is keratins. The epidermis unites subpopulation of migrating cells of treelike types: melanocytes and melanosomes are pigment-producing melanin; Langerhans cells, considered as monokletki derived from bone marrow, and Merkel cells, considered as derivatives of keratinocytes [2].

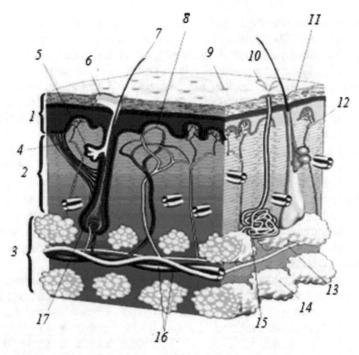

Fig. 1.1 Biological structure of the human skin: *1* epidermic; *2* dermis; *3* hypodermic fatty tissue; *4* muscle, lifting the hair; *5* oil gland, *6* fatty secret; *7* hair ; *8* capillaries; *9* oscule; *10* sweat; *11* keratin (corneal layer); *12* nerve ending; *13* nerve; *14* fat lobule; *15* sweat gland; *16* blood vessels; *17* hair sac

Epidermis is constantly in a state of renovation: the division of basal keratinocytes, some daughter cells move out, and then they separate, go to the upper layers and attach to the corneal layer. Normal human epidermis is renewed in a period which lasts 45–75 days [2]. The leading mechanism in the difficult and multi-stage process directed on the formation of a corneal layer of skin is the formation of a keratin: the main protein in epidermis. In the process of cellular replacement, the epidermis forms the pigment melanin, which is a polymer - granules which are 30–400 nm in diameter [7]. Melanin is produced in melanocytes, containing large number of structural organelles: melanosomes filled with pigment. Melanosomes have a diameter of about 400 nm [7].

Under the epidermis is dense fiber tissue and elastinic tissue, which is called the dermis. Dermis is the main component and volume of the skin. The average thickness of the dermis is about 1500–2000 μm [8]. In the dermis there are the elements of the vascular and nervous systems, the excretory gland. Under the skin is the hypodermis. Hypodermis is a subcutaneous tissue, which is a fatty and connective tissue of varying thickness citebriggman.

The dermis is separated from the epidermal basal membrane and gradually goes into the subcutaneous fatty tissue. The composition of the dense connective tissue in the dermis includes collagen and elastin fibers with a diameter of about 60 nm, packed in bundles of fibers with a diameter about 60 nanometers called - lamels, and an amorphous substance (interfibrilyarny gel), of salt and water. Connective tissue contains widely branched vascular structures of the skin, the nervous network and epithelial glands. Derma is randomly divided into two anatomical areas: papillary (Stratum papillary dermis) and reticular (Stratum reticulare dermis) [9]. More subtle is the so-called papillary dermis, the outer part of the dermal connective tissue that is formed under the epidermis.

Papillary dermis contains more free distribution of elastin and collagen fibers than the reticular layer. Bundles of collagen fibers of papillary dermis are 0.3–3.0 μm in diameter. Papillary dermis also contains lymphatic plexus and blood vessels. The second major part of derma, and the underlying papillary dermis, is called the reticular dermis. Vascular structure of the skin is clearly divided into two systems: a system of vessels that provide nourishment skin and deep, mainly subcutaneous, and arterial and larger venous capillaries which perform the function of heat exchangers of blood with the environment [2].

Biological tissues, such as skin, are optically inhomogeneous absorbing media with an average refractive index greater than that of air, so the boundary between biological object-air part of the radiation is reflected (Fresnel reflection), and the remaining part penetrates the tissue. The skin is characterized by a significant light scattering, i.e., it is highly scattering turbid medium, because it consists of a large number of scattering centers randomly distributed in the volume. The degree of scattering depends on the wavelength of the radiation and the optical properties of biological tissues.

Absorption of light is a physical phenomenon, which characterizes the energy loss when light passes through the biological structure (skin) [10]. The energy of the absorbed light transfered into heat is spent on photochemical reactions or released in the form of radiation luminescence. The absorption spectra of any tissue, including skin, determined by the presence of all biologically important molecules involved in double bonds (chromophores of skin), and containing water in biological tissues. In the epidermis the role of chromophores perform various fragments comprising the amino acids and nucleic acids, absorbing light in the ultra-violet wavelength range [11]. In the visible spectrum one of the most prevailing chromophores of skin is the pigment melanin [2]. The absorption spectrum of melanin has no pronounced absorption bands; however, it effectively absorbs in all spectral regions from 300 to 1200 nm.In the near ultraviolet radiation and visible regions of the spectrum, aside from melanin, the basic skin chromophores are bilirubin, vitamins, flavins, flavin ferments, carotenoids, phycobilins, phytochrome, and others, as well as elastin and collagen fibers [12].

The dermis of the skin include blood vessels, which contain hemoglobin, the absorption spectrum of which significantly affects the absorption spectrum of the skin. The higher the content of blood in the dermis, the greater absorption of its radiation at corresponding to the absorption of blood. Therefore, when calculating the optimal parameters of the radiation blood content in the dermis and the diameter of blood vessels should be considered. In the in vivo biological tissue hemoglobin binds oxygen present in the blood. The absorption spectra of the two forms of hemoglobin are slightly different from each other: oxyhemoglobin has an absorption band near 405 nm (Sore band) and the characteristic double peak absorption in the area of 545–575 nm; deoksigemoglobin strongly absorbs near 430 nm and a weak close to 550 nm [13, 14].

In the infrared region of the spectrum all biomolecules have quite intense vibrational absorption bands.

Starting with $\lambda = 1500$ nm and above, the absorption spectrum of the skin is largely determined by the absorption spectrum of water.

The absorption of subcutaneous fatty tissue is defined as absorption bands of lipids, water, and β- carotene. The main absorption band of fatty tissue lies in the ultraviolet and infrared regions of the spectrum. Skin tissue is characterized by a significant light scattering, as it consists of a large number of randomly distributed scattering centers in volume [15]. Light scattering happens because of fluctuations in the density of scatterer and refractive index fluctuations in the volume of tissue.

The nature of scattering depends on the correlation of the wavelength of the scattered radiation and the size of the light scattering particles, and the ratio of the refractive index of the scattering particle and its environment [16]. Light scattering in media consisting of a large number of particles is significantly different from the scattering of light by individual particles.

This is explained by firstly the interference of the waves scattered by the individual particles with each other and with the incident wave, and second, in many cases, multiple scattering effects are important (reradiation), when the light is scattered by a single particle, others are dissipated again and thirdly, the interaction between the particles does not allow them to consider independent movement.

To account for multiple scattering and absorption of the laser the beam is broadened and attenuated during propagation in the skin. Volume scattering is the cause of a significant proportion of the radiation propagation in the reverse direction (backscattering). Cell membranes, nucleus and organelles are the main scatterers in many biological tissues. The absorbed light is converted into heat, is reradiated as fluorescence or phosphorescence, and spent photobiochemistry reaction.

The absorption spectrum is determined by the type of dominant absorption centers and water content in the tissue. The natural photo of laser radiation of biological tissue is determined by its composition and the absorption coefficient at the wavelength of radiation. The ultraviolet and infrared ($\lambda > 2$ nm) spectral region dominates the absorption and scattering, and the contribution is relatively small and shallow. Light penetrates into the biological tissue, only to one or more cell layers in the short-visible and spectrum of light penetration depth for a typical tissue that is 0.5–2.5 mm. In this case the main role is absorption and thus scattering, which predominates in the reflected radiation from the skin (affects approximately 15–50% of the incident beam). At wavelengths from 600–1500 nm scattering prevails over absorption and penetration depth is increased to 8–10 mm.

Depending on the type of tissue the wavelength of the reflection coefficient can vary widely. Thus, the optical properties of biological tissue are determined by its structure, physiological condition, the level of hydration, homogeneity, specific variance, the nature of the measurements in-vivo - in-vitro and others. The attenuation of the laser beam in biological tissue follows the exponential law. The intensity of the collimated radiation is estimated under Bouguer law. Other important optical parameters of the tissue is the optical depth penetration.

The significant value of the scattering anisotropy of biological tissues and multiple scattering gets a deviation from the Bouguer law. In the description of the effects that occur in the tissues under the influence of radiation, absorption of water is important because it is the main component of most tissues. The human body contains from about 55–65% water. An adult with a body weight of 65 kg contains an average of 40 liters of water, of which about 25 L is inside cells, 15 L are in the extracellular fluids. Water is the primary medium in which many chemical reactions take place and the physical and chemical processes (assimilation, dissimilation, osmosis, diffusion, transport and others) that important for life. In the ultraviolet, visible and near infrared wavelengths the absorption coefficient of water is very small. In these areas, the absorption of tissue determines the absorption spectra of pigments, especially for the skin - the absorption spectra of melanin and blood count (hemoglobin and oxyhemoglobin). Melanin absorption is the most important component of the total absorption of the epidermis and the stratum corneum.

For the calculation of interest the optical density (OD) of the epidermis is needed, which is the result of the following product:

$$OD = \mu_{melanin} \cdot h,$$

where $\mu_{melanin}$ is the absorption coefficient of melanin, h is the thickness of the epidermis.

Optical density depends on the amount of melanin in the basal layer, which depends on many factors, the main one of which is the type of skin. Note that the dermis is very different from the epidermis in the composition and structure. The scattering coefficient of the dermis stronger at shorter wavelengths. Scattering plays a major role in determining the depth of penetration of radiation at different wavelengths in the dermis. Therefore, longer wavelengths penetrate deeper rather than shorter. It is explained by the presence of melanin, which absorbs more shorter wavelengths than long. According to [15] for a sample consisting of the epidermis and dermis, the depth of which is 0.15–0.2 mm (wavelength 632.8 nm) and 0.21–0.4 nm (wavelength 675 nm).

1.3   Structure and Optical Properties of Blood

Blood is one of the most important biological fluids. Blood is the liquid part of the plasma (57% of blood volume) and suspended in its cell (enzymatic) elements (43%). Plasma consist from 90–91% water; 6.5–8.0% are protein molecules and the remaining 2% are low molecular substances. In addition, the blood contains platelets 99% of the blood cells are red blood cells, and 1% are white blood cells and platelets.

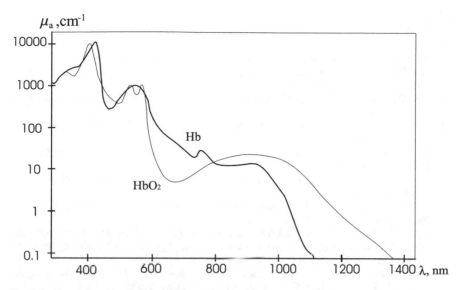

Fig. 1.2 The absorption spectra of hemoglobin and oxyhemoglobin

Red blood cells have a biconcave disk shape with a diameter of about 7μm and a thickness varying from 1–2 μm center to the edges. The cell contains hemoglobin molecules that easily join the oxygen molecules, when they are converted into oxyhemoglobin. Accordingly, we have different venous and arterial blood. The hematocrit is volume percentage of red blood cells in whole blood.

The most important parameter is also the oxygen saturation (OS), defined as the ratio of oxygenated hemoglobin to total hemoglobin. The absorption of the blood is determined mainly by water absorption, hemoglobin and oxyhemoglobin. The absorption spectra of these pigments is shown in Fig. 1.2 [17]. If the hematocrit increases, this means that the number of red blood cells is increasing and there is an increase in the scattering. At higher hematocrit $H > 0.5$ erythrocytes stick together, forming a homogeneous mass absorbed by hemoglobin and scattering occurs on the plasma cavity located between the masses of red blood cells. This section contains the optical parameters of the biological structures without their temperature dependences. Note that with increasing temperature the optical characteristics of tissues and their components will change.

Now consider the basic principles of mathematical models to calculate the interaction of laser radiation with a turbid medium. One example of such an environment is human biological tissue. Biological tissue is a multilayer medium containing various inclusions, such as, for example, blood vessels, in which the blood moves. Consider the main approaches in the theory of mathematical models that describe the interaction of laser radiation with multi-layered turbid media.

References

1. T.T. Berezov, B.F. Korovkin, *Biological Chemistry* (Medicine, Moscow, 1990)
2. E.A. Pylypenko, The reflectivity and fluorescence spectroscopy of human skin in vivo: Ph.D. dissertation. Saratov: Saratov State University, 1998
3. G.D. Weinstein, R.J. Boucek, Collagen and elastin of human dermis. J. Investig. Dermatol. **35**, 227–229 (1960)
4. G.F. Odland, The morphology of the attachment between the dermis and the epidermis. Anat. Rec. **108**, 339–413 (1950)
5. S.L. Jacques, The role of skin optics in diagnostic and therapeutic uses of lasers, in *Lasers in Dermatology. B*. (Springer, Berlin, 1991), pp. 1–21
6. A.M. AM Chernuha, E.P. Frolov (eds.), *The skin (structure, function, general pathology, therapy)* (Moscow, 1982)
7. N. Kollias, R.M. Sayer, L. Zeise, M.R. Chedekel, Photoprotection by melanin. J. Photochem. Photobiol. B. **9**, 135–160 (1991)
8. I.V. Meglinski, Simulation of the reflectance spectra of optical radiation from a randomly inhomogeneous multilayer strongly scattering and absorbing light environments using the Monte Carlo. Quantum Electron. **31**(12), 1101–1107 (2001)
9. G.F. Odland, Structure of the skin, in *Physiology, Biochemistry, and Molecular Biology of the skin*, ed. by L.A. Goldsmith (Univ. Press, Oxford, 1991), pp. 3–62
10. V.V. Tuchin, Light scattering study of tissues. Successes Phys. Sci. **167**, 517–539 (1997)
11. U.A. Vladimirov, A.Y. Potapenko, *Physical and Chemical Basis of Photo-biological Processes* (Moscow, 1989)
12. S.R. Utz, J. Barth, P. Knuschke, YuP Sinichkin, Fluorescence spectroscopy of human skin. Proc. SPIE **2081**, 48–57 (1993)
13. P.H. Andersen, P. Bjerring, Spectral reflectance of human skin in vivo Photodermatol. Photoimmunol. Photomed. **7**, 5–12 (1990)
14. R.R. Anderson, J.A. Parrish, K.F. Jaenicke, Optical properties of human skin, in *The Science Photomedicine*, ed. by J.D. Rogan, J.A. Parrish (Plenum Press, New York, 1982), pp. 147–194
15. W.-F. Cheong, S.A. Prahl, A.J. Welch, A review of the optical properties of biological tissue. IEEE J. Quantum Electr. **26**(12), 2166–2185 (1990)
16. Bohren, Huffman, *Absorption and Scattering of Light by Small Particles. M*. (1986)
17. K. Johnson, A. Guy, Impact of non-ionizing electromagnetic radiation on biological systems and the environment. Proc. IEEE **60**(6), 49–79 (1972)

Chapter 2
Overview of Theoretical Approaches to the Analysis of Light Scattering

Abstract We propose methods of light scattering for the quantitative study of the optical characteristics of the tissue, and the results of theoretical and experimental studies of photon transport in biological tissues.

2.1 Introduction

One important aspect of the development of modern medicine is the early detection of diseases. To solve this problem it is necessary to select the most informative indicators of the functioning of the organism, such measurements are the results of the analysis of peripheral blood. As we know the blood is made up of the following elements: white blood cells, red blood cells and platelets, the optical properties of biological objects help to solve a number of problems for the diagnosis of various pathological processes in the body. In medical diagnostic methods are divided into ≪invasive≫ and ≪non-invasive≫. Invasive methods assume such an action on the prototype system, such as is going through some changes (X-rays) in the organs or tissues. Non-invasive methods are methods in which information about an object is obtained without disturbance of the internal structure of the body. Considering that the classification is from a physical viewpoint we can say that completely non-invasive methods do not exist in the very nature of the measurement procedure. It is correct to describe all methods of diagnosis to some degree as perturbation, introduced into the prototype system, and those where a disturbance is minimal considered non-invasive methods. As such, a new area of diagnostics was now actively developed, which uses optical techniques. With the optical point of view of biological tissue (including bioliquid: blood, lymph, etc.) it can be divided into two broad classes:

1. strongly scattering (skin, vascular wall, blood);
2. weakly scattering (tissue of the eye: cornea, crystalline lens) [1, 2].

This chapter describes the use of light scattering techniques for a highly scattering biological structure.

To develop new methods of laser biomedical diagnostics one must study in detailed the peculiarities of the process of light propagation in biological tissues, as theoretical studies improve understanding of the optical measurements, reliability and usefulness

© Springer International Publishing AG, part of Springer Nature 2018

K. Kulikov and T. Koshlan, *Laser Interaction with Heterogeneous Biological Tissue*, Biological and Medical Physics, Biomedical Engineering, https://doi.org/10.1007/978-3-319-94114-1_2

of optical technologies. The use of light scattering techniques for biological particles was developed in the articles based on the Mie theory for single, two and three particles [3].

With the help of a two-layer model of the sphere one hase described light-scattering properties of suspensions of erythrocyte [4, 5].

Note that the first articles involving the exact theory of electromagnetic waves of a two-layer sphere considered a model of biological particles [6, 7]:

1. the refractive index of the particle and its shape;
2. parts inside the particle, i.e, antrum, small inhomogeneities [8, 9].

These studies have identified the optical properties of the typical representatives of biological particles: angular dependence of the red blood cells, in reducing platelets, which are associated with changes in cell shape. Effects of aggregation and dispersion.

As also the different characteristics of the native cell were evaluated [10, 11].

However, elucidating the physical mechanisms of living systems and the development of pathological changes requires new methods for studying living matter and the manipulation of biological structures. Thus, in the optics of scattering media there are three main directions.

The first direction. This direction is connected with the solution of diffraction problems for the individual plates and the linking characteristics of the absorption and scattering of the optical geometry and structural parameters of the particles. In this field of research a number of new methods and algorithms has been developed to obtain quantitative results for a broad class of sizes, shapes, structures, and optical parameters of the particles.

The second direction of light scattering theory is associated with the equation of radiative transfer. This equation uses the photometric values and phenomenological characteristics of the environment, namely, the scattering coefficient, absorption and scattering function of volume. In the multiple scattering theory of transport phenomenology is taken into account and based on the law of conservation of energy and the concept of radiation intensity.

The third direction of scattering theory, electrodynamics consideres statistically inhomogeneous media. This approach takes into account the multiple scattering of waves (MSW) in the discrete or continuous irregularities and the vector nature of the electromagnetic field. The theory of multiple scattering of waves is based on simple physical principles. First, we assume that we know the spatial configuration of all the particles and their statistical properties. Secondly, it is assumed that we know scattering operator of a single particle, which describes the scattered field for a given excited field. Note that the excited field is the sum of the incident field and the field of multiple scattering from all other particles, because we consider the electrodynamic of system interaction with multipole oscillators. Thus,to find an excited field with all possible orders of scattering from all the interacting particles, is the main difficulty of the theory.

Various versions of this theory differ primarily only in the ways of approximate calculation of the excited field with the statistical properties of the ensemble, which describes the spatial configuration and the optical properties of lenses If the excited field is found, further analysis concerns the calculation from the scattered fields of individual particles and the addition of these fields with phase shifts. Since we are considering random fields, calculating the observed photometric is needed to use the correlation analysis. In the theory of multiple scattering of waves the theory of coherent radiation propagation in a medium close-packed lenses has been developed in detail, the main result of which is the output of the dispersion equation for the effective wave number describing the propagation of a coherent field in a medium different from the wave number of free space. This dispersion equation takes into account the optical properties of the scatterers and the statistical properties of their spatial location. Note that in the derivation of the dispersion equation simplifying assumptions are made. For example, the use of quasi-crystalline approximation to decouple the infinite chain of equations of multiple scattering, and to describe the pairing correlations in the positions of the particles using the Percus-Yevick. A fundamental feature of the theory of MSW is that the optical properties of interacting particles differ from those characteristics which are obtained by solving the scattering problem for an isolated particle. For example, the extinction cross-section of the particles in the cluster do not coincide with the usual calculation of Mie theory. Even in the simplest case of two completely identical spheres in the contact cross section of each particle depends on the orientation of the bisfery in relation to the incident plane wave. Effects of this type are said to be "collective" or "cooperative" effects of the scattering of interacting particles. In general, the cooperative effects of multiple scattering are the two components and their calculation is rather complex. However, for biological systems the situation is simplified by the fact that the optical properties of interacting particles are usually not much different from those of the environment. The analysis of the conditions for the applicability of a specific version of the theory of light scattering is a nontrivial problem, which requires taking into account the coherence properties of the incident light, the size, concentration, and optical properties of the particles, the time of stability of the microstructure of the medium (i.e., the characteristic relaxation times of fluctuations), the geometric parameters of the scattering sample, the characteristics of the photodetector, etc. Note that in this paper we consider only the first two of the main approaches in the theory of scattering for highly scattering tissue.

2.2 Optical Properties of Tissues with Multiple Scattering

In this section we consider light scattering methods for the quantitative study of the optical characteristics of the tissue, and the results of theoretical and experimental studies of photon transport in biological tissues. The theoretical analysis is based on a stationary or non-stationary radiation transfer theory for strongly scattering media, as well as numerical Monte-Carlo method, which is used to solve the problems of scattering in multilayered biological tissues with complex boundary conditions.

2.3   Stationary Theory of Radiative Transfer

Transport theory the theory of radiative transfer was developed by Schuster in 1903 [12]. Transport theory does not include diffraction effects. In the classical theory of radiative transfer, considering the wave field as a combination of incoherent radiation beams, the basic concept is the radiation intensity (or brightness) $I(\mathbf{r}, \mathbf{s})$, which determines the average energy flux dP through the surface element $d\alpha$ that is concentrated in a solid angle $d\Omega$ near the direction $\mathbf{s}$ of the frequency interval $(\nu, \nu + d\nu)$:

$$dP = I(\mathbf{r}, \mathbf{s}) \cos\theta d\alpha d\Omega d\nu \qquad (2.1)$$

Stationary equation of radiative transfer theory for monochromatic light has the form [13]:

$$\frac{\partial I(\mathbf{r}, \mathbf{s})}{\partial \mathbf{s}} = -\mu_t I(\mathbf{r}, \mathbf{s}) + \frac{\mu_s}{4\pi} \int_{4\pi} I(\mathbf{r}, \mathbf{s}')p(\mathbf{s}, \mathbf{s}')d\Omega', \qquad (2.2)$$

where $I(\mathbf{r}, \mathbf{s})$ is ray intensity at the point $\mathbf{r}$ in the direction $\mathbf{s}$, $p(\mathbf{s}, \mathbf{s}')$ is phase function of the scattering, $d\Omega$ is unit solid angle in the direction $\mathbf{s}'$, μ_s is scattering coefficient, $\mu_t = \mu_a + \mu_s$ is coefficient of the total interaction, μ_a is coefficient of absorption. We assume that there are no light sources inside the medium.

The boundary condition for the equation (2.2) is:

$$I(\mathbf{r}, \mathbf{s})|_{(\mathbf{sn})<0} = I_Q(\mathbf{r}, \mathbf{s}) + RI(\mathbf{r}, \mathbf{s})|_{(\mathbf{sn})>0}, \mathbf{r} \in \partial\Gamma, \qquad (2.3)$$

where $I_Q(\mathbf{r}, \mathbf{s})$ is boundary distribution of radiation intensity generated by external sources, $\mathbf{n}$ is outward normal to the $\partial\Gamma$ at $\mathbf{r}$, R is the operator of reflection.

The phase function $p(\mathbf{s}, \mathbf{s}')$ describes the scattering properties of the medium and is the probability density function of the scattering of photons in the direction of the $\mathbf{s}'$ which move in the direction of $\mathbf{s}$. The phase function $p(\mathbf{s}, \mathbf{s}')$ can be defined as a table form, derived from measurements or represented by an analytic expression.

In many practical cases, the phase function is well approximated using the empirical Henie-Greenstein function

$$p(\theta) = \frac{1}{4\pi} \frac{1 - g^2}{(1 + g^2 - 2g\cos(\theta))^{3/2}},$$

where g is the scattering anisotropy factor,

$$\frac{\partial I_{ri}(\mathbf{r}, \mathbf{s})}{\partial \mathbf{s}} = -\mu_t I_{ri}(\mathbf{r}, \mathbf{s}), \qquad (2.4)$$

I_{ri} is the weakened incident intensity. Note that the expression (2.4) coincides with the Bouguer law for the scattering medium. This means that for the weak incident intensity in the transport theory Bouguer law is valid for all optical thicknesses.

The total intensity is determined as

$$I(\mathbf{r}, \mathbf{s}) = I_{ri}(\mathbf{r}, \mathbf{s}) + I_d(\mathbf{r}, \mathbf{s}), \tag{2.5}$$

and satisfies the equation (2.2) while diffuse intensity is determined by the equation

$$\frac{\partial I_d(\mathbf{r}, \mathbf{s})}{\partial \mathbf{s}} = -\mu_t I_d(\mathbf{r}, \mathbf{s}) + \frac{\mu_s}{4\pi} \int_{4\pi} I_d(\mathbf{r}, \mathbf{s}') p(\mathbf{s}, \mathbf{s}') d\Omega' + \varepsilon_{ri}(\mathbf{r}, \mathbf{s}), \tag{2.6}$$

where $\varepsilon_{ri}(\mathbf{r}, \mathbf{s})$ is the function of the equivalent source.

The scalar equation (2.2) is used in optics to describe light in cases where polarization effects can be ignored.

Exact solutions of the transport equation and the integral equation for the radiation intensity are obtained only for a small number of special cases. Examples of this kind for which solutions are found and stored in a suitable form for the calculations are coplanar problems and problems with isotropic scattering.

We consider several approximations that are often used in optics of biosystems.

2.4 Approximate Methods for Solving the Transport Equation

First order approximation. In the case of weak scattering the scattering medium is sparse, and the scattering volume is not large, solving the transport equation can be obtained by iteration.

In the first approximation, the iterative solution of the radiative transfer equation produces a result, known as a first order approximation transfer theory [14]. In this approach it is assumed that the total intensity incident on the particles is approximately equal to the incident intensity weakened, which is known. Consequently the solution to the first-order approximation of the form [14] is:

$$I(\mathbf{r}, \mathbf{s}) = I_{ri}(\mathbf{r}, \mathbf{s}) + I_d(\mathbf{r}, \mathbf{s}) \tag{2.7}$$

$$I_d(\mathbf{r}, \mathbf{s}) = \int_0^s \exp[-(\tau - \tau_1)] \left[\frac{\mu_s}{4\pi} \int_{4\pi} I_{ri} d(\mathbf{r}, \mathbf{s}') p(\mathbf{s}, \mathbf{s}') d\Omega' \right] d\mathbf{s}', \tag{2.8}$$

where I_{ri} is the weakened incident intensity, I_d is the diffuse intensity, τ, τ_1 are optical paths,

$$\tau = \int_0^s \rho \mu_t ds, \tau_1 = \int_0^{s_1} \rho \mu_t ds$$

and ρ is the total number of particles per unit volume. Note that the solution to the first-order approximation is valid for optically thin and weakly scattering media ($\tau < 1$, $\Lambda < 0.5$) when the intensity of the transmitted wave (coherent component) is described by the Bouguer law. In the case of a very sharp incident beam (eg, laser light) first order approximation is valid for more dense tissues ($\tau < 1$, $\Lambda < 0.5$), where $\Lambda = \mu_s/\mu_t$ — is the single scattering albedo.

Diffusion approximation. This approach suggests that diffusion intensity encounters many particles and disperses them nearly uniformly in all directions, so it is the almost isotropic angular distribution [12]. Diffused illumination components can be represented in the form of spherical harmonics of Legendre polynomial [15]. We consider only first two terms in the expansion in the series, then we have the diffusion approximation, which is written as

$$L_s(\mathbf{r}, \mathbf{s}) = \frac{1}{4\pi} \int_{4\pi} L_s(\mathbf{r}, \mathbf{s}) d\Omega + \frac{3}{4\pi} \int_{4\pi} L_s(\mathbf{r}, \mathbf{s}')\mathbf{s}' \cdot \mathbf{s} d\Omega =$$

$$= L_0(\mathbf{r}) + \frac{3}{4\pi} F(\mathbf{r}) \cdot \mathbf{s}, \qquad (2.9)$$

where $L_0(\mathbf{r})$ is the indexaverage diffuse intensity, $F(\mathbf{r})$ is the diffuse flux vector oriented along the direction of the unit vector $\mathbf{s}$. The first of these equations expresses Fick law (power density is proportional to the gradient of light), which describes the increase or decrease of the power flux density due to absorption and scattering of collimated and diffuse components:

$$F(\mathbf{r}) = -\frac{1}{3\mu_\sigma}\nabla\varphi_s(\mathbf{r}) + \frac{\mu_s g}{\mu_\sigma}E(\mathbf{r}, \mathbf{s_0}) \cdot \mathbf{s_0}, \qquad (2.10)$$

where $\mu_\sigma = \mu_a + (1 - g)\mu_s$ is transport damping factor. The second equation is described in the following expression:

$$\nabla \cdot F(\mathbf{r}) = -\mu_a \varphi_s(\mathbf{r}) + \mu_s E(\mathbf{r}, \mathbf{s_0}) \qquad (2.11)$$

Thus, in the stationary case, the transport equation in the diffusion approximation can be written as [15]:

$$\nabla^2 \varphi_s(\mathbf{r}) - 3\mu_a\mu_\sigma\varphi_s(\mathbf{r}) + 3\mu_s\mu_\sigma E(\mathbf{r}, \mathbf{s_0}) - 3\mu_s g \cdot \nabla(E(\mathbf{r}, \mathbf{s_0})\mathbf{s_0}) = 0 \qquad (2.12)$$

Biological tissues scatter light mainly in the forward direction. As a result, the diffusion approximation is not always a good approximation of the theory of radiation transport near sources or boundaries. To improve the situation we include the delta function in the definition of the phase function [15]:

$$p(\mathbf{s}, \mathbf{s}') = (1 - f)p'(\mathbf{s}, \mathbf{s}') + f\delta(1 - \mathbf{s} \cdot \mathbf{s}')\frac{1}{2\pi}. \qquad (2.13)$$

This representation is called the Delta-Eddington approximation.

The diffusion equation in this case can be written using the new variables:

$$\mu'_t = \mu_a + \mu'_t, \, \mu'_s = \mu_s(1 - f), \, p'(\mathbf{s}, \mathbf{s}'), \, f = g^2, \, g' = \frac{g}{g + 1}$$

These coefficients correspond to a phase function of type Henie-Greenstein approximation. Transformation $p \to p'$ (p' is new phase function) is a mathematical transformation. Changes occur in the source region and borders, and this is especially important for a strong forward scattering.

Delta-Eddington approximate reduces the degree scattering direction ($g' < g$).

The boundary condition for the solution of the transport equation can be written as:

$$\int_{2\pi} L_s(\mathbf{r}, \xi)(\xi \cdot \mathbf{n})d\Omega = 0, \tag{2.14}$$

where $\mathbf{n}$ is the unit normal vector.

The boundary condition for solving the transport equation in the diffusion approximation at the boundaries with air can be written as [15]:

$$\frac{1 - r_{21}}{1 + r_{21}} \cdot \frac{\varphi_s(\mathbf{r})}{2} + \frac{\mu_s g}{\mu_\sigma} E(\mathbf{r}, \xi_0)\mathbf{n} - \frac{1}{3\mu_\sigma}\nabla\varphi_s(\mathbf{r})\mathbf{n} = 0 \tag{2.15}$$

where r_{21} is reflection coefficient at the air-biological tissue.

It is necessary to distinguish three types of boundaries with air which are as follows: the higher boundary to which the radiation drops, the side boundaries and the lower boundary of the tissue. For these kinds of boundaries the reflection coefficients are different. For the upper boundary, through which radiation from the air enters the scattering medium, this coefficient has the form [16]

$$r_{21} = 1 - \left(\frac{1}{n_2}\right)^2$$

for the lower and side boundaries, through which radiation from the environment goes into the air the factor has the following form:

$$r_{21} = \frac{\cos^2\theta_c + \cos^3\theta_c}{2 - \cos^2\theta_c + \cos^3\theta_c},$$

where

$$\theta_c = \arcsin\left(\frac{1}{n_2}\right).$$

At internal borders the given condition is equality flow.

The diffusion theory is a good approximation in cases where the anisotropy of scattering is small ($g \leq 0.1$) and scattering albedo is large $\Lambda \longrightarrow 1$. For many tissues the scattering anisotropy factor is $g \approx 0.6 - 0.9$, and in some cases, for example blood, can reach values of 0.990–0.999 [17]. This substantially restricts applicability of the diffusion approximation. Several papers were devoted to the study of the accuracy of the diffusion approximation [18]. Based on the comparison of solutions of the diffusion equation with the results of the Monte-Carlo simulation[19] (see below) one concluded that the diffusion approximation may be a solution on some orders from the truth. In optics the tissues simpler methods have been found for solving the transport equation, such as the two-flux Kubelka-Munk model, three-, four-and seven-flux models [12].

Two-and multiflux approximation. This theory is based on the model of the two light beams propagating in the forward and backward directions. The main assumption of this theory is that the radiation intensity is diffuse. Inside the tissue diffuse flux is divided into two parts: L_1 the flow in the direction of incident radiation and flux scattered back L_2. For the absorption and scattering of diffuse radiation we introduce two Kubelka-Munk coefficients: A_{KM} and S_{KM}.

We have two differential equations

$$\frac{dL_1}{dz} = -S_{KM}L_1 - A_{KM}L_1 + S_{KM}L_2$$

$$\frac{dL_2}{dz} = -S_{KM}L_2 - A_{KM}L_2 + S_{KM}L_1,$$

where z is the average direction of the incident radiation.

Coefficients A_{KM} and S_{KM} values μ_a and μ_s are written as follows [12]: $A_{KM} = 2\mu_a, S_{KM} = \mu_s$.

The Kubelka-Munk theory is a special case of multiflux theory, where the transport equation is transformed into a matrix differential equation which takes into account the intensity of the radiation in the direction of many of the individual solid angles.

The two-flux theory is not applicable to describe the incident on a medium collimated beam. In this case, we use the four-flux theory. The Four-flux theory [12] takes into account two counter diffuse flux as the Kubelka-Munk theory, as well as two collimated laser beams, the external incident and reflected from the back surface of the sample The Seven- flux theory is a three-dimensional representation of the incident laser beam and the scattering of radiation in a semi-infinite medium [20]. Note that the Kubelka-Munk theory can only be applied to a one-dimensional geometry of the system. The numerical approximation of the transport equation can be obtained by the Monte-Carlo method.

Monte-Carlo method. The general scheme of the Monte-Carlo method is based on the central limit theorem of probability theory. General properties of the Monte-Carlo method:

- absolute convergence of the solution is of order $\frac{1}{N}$;
- independence of the error on the number of tests is of order approximate $\frac{1}{\sqrt{N}}$
- the main method reducing the error is the maximum variance reduction;

- the error does not affect the dimension of the problem;
- simple structure of the computational algorithm;

From the viewpoint of the solutions of the equation for radiative transfer the Monte-Carlo method is a computer simulation of the random motion of N photons [14]. To obtain reasonable approximation to one have consider a large number of photons because the accuracy of the results is proportional to $\sqrt{N}$. The main idea of the method is the registration of effects of absorption and scattering throughout the optical path of a photon through a non-transparent environment. The distance between two collisions is chosen from a logarithmic distribution, using a random number generated by a computer. To take into account absorption, each photon is assigned a weight.

If there is scattering, a new direction propagation is chosen according to phase function and other random number. This procedure is repeated as long as the photon does not come out of the considered volume or the weight reaches a certain value. The Monte-Carlo method includes five main steps: generation of the source photon trajectory, absorption, destruction, registration [14].

1. Generation of photon source. The photons are generated on the surface of the medium. Their spatial and angular distribution corresponds to the distribution of the incident radiation (for example, a Gaussian beam).
2. The generation of the trajectory. After generation of a photon the distance to the first collision is determined. We expect that the absorbing and scattering particles are randomly distributed in opaque medium. Then, the value of free path is $1/\rho\sigma_x$, where ρ is particle number density and σ_x is the scattering cross-section. Random number $0 < \xi_1 < 1$ is generated by computer and the distance to the next collision $L(\xi_1)$ is calculated from the expression

$$L(\xi_1) = -\frac{\ln \xi_1}{\rho\sigma_x}.$$

   Since

$$\int_0^1 \ln \xi_1 d\xi_1 = -1,$$

   average quantity $L(\xi_1)$ is $1/\rho\sigma_x$. From this we obtain a scattering point. The scattering angle is determined by the second random number ξ_2 in accordance with the phase functions, such as Henie-Greenstein function. The polar angle Φ is determined by the expression $\Phi = 2\pi\xi_3$, where ξ_3 is a third random number between 0 and 1.
3. Absorption. To take into account the absorption, we assigned weight to each photon. At the point of entry to the opaque medium, the weight of a photon is equal to 1. The weight decreases by absorption in accordance with the expression $\exp[-\mu_a L(\xi_1)]$. As an alternative to assigning weights a fourth random number ξ_4 can be added ($0 < \xi_4 < 1$).

We assume that the scattering takes place only when, $\xi_4 < a$ is where optical albedo,

$$a = \frac{\mu_s}{\mu_a + \mu_s}.$$

if $\xi_4 < a$ photon is absorbed, which is analogous to step 4.

4. Destruction. This step is used only when assigning a weight to each photon in step 3. When the weight reaches a certain value, the photon is eliminated. Then a new photon emits and the program continues with step 1.
5. Registration. After repeating steps 1–4 for a sufficient number of photons, a map of the trajectories is calculated and accumulated in the computer. Thus, it may be obtained by a statistical report on portions of the incident photons absorbed by the medium, and the spatial and angular distribution of the photons emerging from it.

We consider one of the variants of the construction algorithm of the Monte-Carlo method. The modeling medium is defined by the following parameters: L_{ave} is the thickness, μ_s is the scattering coefficient and μ_a is the absorption coefficient, g is the cosine of the scattering angle, n is the relative refractive index.

The incident impulse consists of one million photons within the medium along the z-axis perpendicular to the surface (x, y) at the point $(0, 0, 0)$. Calculations are made in a three-dimensional Cartesian coordinate system. After entry of the photon the mean free path of a photon in the medium, and the scattering angles θ and φ are determined. The scattering angle $p(\theta)$ is defined by the scattering phase function. In the general case $p(s, s') = p(\theta)p(\varphi)$ where s is incident direction, s' is scattering direction of photon. Note, particles of medium are spherically symmetrical particles, when we have absorption and scattering. This approximation is used in similar cases, and based on the fact that in the process of passage through a medium with strong scattering of a photon interacts with particles from different angles. We can therefore use the average of the scattering indicatrix.

Thus, if you use this approach, we have $p(\varphi) = \frac{1}{2\pi}$. In the case of tissue with strong scattering as a function of the phase of the scattering phase function $p(\theta)$ Henie-Greenstein can be applied, from which we obtain an expression for the angle θ:

$$\theta = \cos^{-1}\left[\frac{1 + g^2 - \left[\frac{1-g^2}{1+g^2-2g\,Random}\right]^2}{2g}\right],$$

where Random is random number uniformly distributed in the range (0,1). At each step θ angle is relative to the ≪old≫ direction of propagation, the angle φ is in a plane perpendicular to the ≪new≫ direction of movement.

The free path of photon is:

$$p(L) = \left[\frac{1}{l_{ph}}\right]^{\frac{l}{l_{ph}}},$$

where mean free path of photon is

$$l_{ph} = \frac{1}{\mu_a + \mu_s}$$

Since

$$\int_0^\infty p(L)dL = 1.$$

For the calculation of the mean free path we take random number $\xi \in (0, 1)$:

$$\xi = \int_0^L p(l)dl.$$

The number ξ, which is uniformly distributed in the interval $(0, 1)$, is given as computer generated random number.

Thus, the free path of a photon is:

$$L = -l_{ph} \ln(l - \xi).$$

After that one models the interaction of a photon with a particle of the medium, which can be either absorbing or scattering center. The probability of photon scattering on the particle is

$$p_s = \frac{\mu_s}{\mu_s} + \mu_a,$$

The probability of absorption is:

$$p_a = \frac{\mu_a}{\mu_s} + \mu_a = 1 - p_s.$$

If the generator produces a random number in the range $(0, p)$, then the photon is considered to be scattered, otherwise it is absorbed. The total layer of the medium along the z-axis virtually divided into a number of thinner layers having equal thickness to which data arrays correspond. In each array the number of absorbed or scattered photons is recorded. Thus, the spatial resolution of the depth of the sample is

$$\frac{1}{L_{ave}}.$$

If the photon is scattered, its new direction and coordinates are calculated with the following formulas::

$$x = x_0 + L \sin \theta \cos \varphi,$$

$$y = y_0 + L \sin \theta \sin \varphi,$$

$$z = z_0 + L \cos \theta,$$

where x_0, y_0, z_0 are the ≪old≫ coordinates of photon. If the photon is absorbed, then we start the next one. Next, all the coordinates are translated in the original coordinates. Calculation continues as long as the photon is not absorbed or leaves the detector. At the boundaries of the medium-to-air total internal reflection is:

$$\theta = \sin^{-1} \left(\frac{1}{n} \right),$$

where n is refractive index of medium. Note that the use of Monte-Carlo method is based on the use of macroscopic optical properties of the medium which are assumed to be homogeneous within small volumes of tissue and simulation by the Monte-Carlo method does not account for details of the energy distribution of radiation inside an individual cell.

2.5   The Nonstationary Theory of Radiative Transfer

Using the nonstationary transfer theory, we can analyze the response time of the scattering tissue [21]. This analysis is important for justification of noninvasive optical methods using measurement reflection or transmission of tissue with a time resolution [1, 22, 23]. The nonstationary equation of radiative transfer theory is [21]:

$$\frac{\partial I(\mathbf{r}, \mathbf{s}, t)}{\partial \mathbf{s}} + t_2 \frac{\partial I(\mathbf{r}, \mathbf{s}, t)}{\partial t} = -\mu_t I(\mathbf{r}, \mathbf{s}, t) +$$

$$+ \frac{\mu_s}{4\pi} \int_{4\pi} \left[\int_{-\infty}^{t} I(\mathbf{r}, \mathbf{s}', t) f(t, t') dt \right] p(\mathbf{s}, \mathbf{s}') d\Omega', \qquad (2.16)$$

where t is time, t_2 is the average time between the interactions,

$$f(t, t') = \frac{1}{t_1} \exp \left(-\frac{t - t'}{t_1} \right),$$

t_1 is the first moment of the distribution function $f(t, t')$ and it means the duration of the individual act of scattering, $t \to 0$, $f(t, t') \to \delta(t - t')$, $I(\mathbf{r}, \mathbf{s}, t)$ is ray intensity. Equation (2.16) satisfies the boundary conditions (2.3) for $(\mathbf{r}, \mathbf{s}) \to (\mathbf{r}, \mathbf{s}, t)$. If the direction $I(\mathbf{r}, \mathbf{s}, t)$ is insignificant compared to the isotropic component, then (2.16) is transformed into a diffusion equation [12, 23]

$$\left(\nabla^2 - c\mu_a D^{-1} - D^{-1} \frac{\partial}{\partial t} \right) U(\mathbf{r}, t) = -Q(\mathbf{r}, t), \qquad (2.17)$$

if $\mu_a = 0$ the diffusion equation is equivalent to the heat conduction equation.

The solution of the diffusion equation (2.17) for media with constrained geometry requires that the source function of the boundary conditions were set as follows [21, 24]:

1. During the sensing medium directional beam radiation source of the diffuse component is not localized on the surface of the medium, but at a certain depth.
2. Boundary condition for the classical diffusion problem can be written as

$$U(\mathbf{r}, t)|_{\mathbf{r}=\Omega} = 0,$$

where Ω is surface bounding the region of space where the diffusion takes place.

In the case of diffusion of radiation, this condition must be modified to account for the influence of the light reflection at the boundary.

Solution of the diffusion of radiation in bounded regions space can be obtained using standard techniques of solving boundary-value problems, for example in the areas in the form of a half-space [25]. Note that the important question is of the influence of absorption on the transport properties of the scattering medium. The diffusion theory of radiation diffusion coefficient is

$$D = \frac{c}{3(\mu_a + (1 - g)\mu_s)}.$$

However, in [23, 25] it is written that a better match between the experiment and the diffusion theory achieved if D is of the form

$$D = \frac{c}{3(1 - g)\mu_s},$$

This allows us to analyze the statistics of the optical paths in the case of an absorbing medium by calculating the probability density $p(s)$ for non-absorbing medium with the specified μ_s and g.

We note that there have been various attempts to modify the diffusion approximation in order to obtain an analytical description of the radiative transfer near the scattering medium, and also for cases of strong absorption and anisotropic scattering. Thus, in [26] the description of the radiative transfer is examined using a three-dimensional telegraph equation.

2.6 Methods for Measuring Optical Parameters of Biological Tissues

To measure optical parameters of biological tissues (absorption coefficient, scattering coefficient) different methods are used. These methods can be divided into two classes: direct and indirect. The direct methods are the methods which are based on the basic concepts and definitions, such as Bouguer law:

$$I(z) = (1 - R)I_o \exp(-\mu_t z), \qquad (2.18)$$

where R is the reflection coefficient, I_o is the intensity of the incident light, μ_t is the absorption coefficient and z is depth.

The measured parameters are the scattering function or lighting inside the volume of the medium. The advantages of these methods include the comparative simplicity of analytical expressions that are used in data processing. The disadvantages of direct methods are related to the need of strict implementation of the experimental conditions, the relevant models: the single scattering for thin samples, the refraction of light on the edges of of the cuvette.

Indirect methods involve solving the inverse scattering problem using specific theoretical model of light propagation in the medium. Indirect methods are divided into iterative and noniterative. Noniterative methods use the equations in which the optical properties are determined by the parameters associated with the measured values. Note, in case in vitro measurements of the parameters of samples of biological tissues we can use method of integrating the two spheres combined with measurements of the collimated transmission.

It consists of consistent or simultaneous measurement of three parameters: the collimated transmission, diffuse transmission T_d and diffuse reflection R_d. For determining the optical parameters of the tissue from these measurements one can use various theoretical equations or numerical methods (two-and multi-flux model, inverse Monte-Carlo method), which establish the relationship between the absorption coefficient, the scattering coefficient with the measured parameters. In the simplest case, we can take a two-flux the Kubelka-Munk model [27]:

$$S = \ln\left(\frac{1 - R_d(a - b)}{T_d}\right); K = S(a - 1);$$

$$a = \frac{1 - T_d^2 + R_d^2}{2R_d}; b = (a^2 - 1)^{1/2};$$

$$K = 2\mu_a; S = \frac{3}{4}\mu_s(1 - g) - \frac{1}{4}\mu_a;$$

$$\mu_t = \mu_a + \mu_s; \mu_s' = \mu_s(1 - g) > \mu_a.$$

Determination μ_t of collimated transmission measurements on the basis of (2.18) allows us, with the help of experimental data T_d, R_d to find all three of the optical parameters of tissue: μ_a, μ_s, g. The Kubelka-Munk model, three-, four-, and seven-flux [20], [27]−[28] are the basis of of indirect noniterative methods.

2.7 Methods for Solving Inverse Problems of Scattering Theory

One method of solving the inverse scattering problem is the inversion method of adding-doubling [28]. The inversion method of adding-doubling includes the following steps:

1. assignment of expected optical parameters;
2. calculation of reflection and transmission using the method of adding-doubling;
3. comparison of the calculated values of the reflection and transmission with the measured;
4. repetition of the procedure to obtain coherent data with a given precision.

The method is used with the following assumptions: the distribution of light is independent of time, the samples have homogeneous optical properties, the geometry of the samples is an infinite plane-parallel layer final thickness, tissue has a homogeneous index of refraction, internal reflection at the boundaries is described by Fresnel law and the light is not polarized The inversion method of adding-doubling was successfully used in finding the optical parameters of the dermis [29].

2.8 Resume

This chapter describes methods for modeling the interaction of light with biological tissue: the diffusion approximation, the theory of radiative transfer, various multi-flux theories and the Monte-Carlo method.

Note the most significant limitations and disadvantages of these methods:

1. The theory of radiative transfer is true for sufficiently distant scatterers.
2. The diffusion approximation can not be applied at a wavelength $\lambda = 0.514\,\mu$m. It also is not applicable near the surface of the object at the input of the light beam, where single scattering is predominant.
3. A major shortcoming of the Monte-Carlo method is that in order to obtain precise results with help program must be passed a large number of photons.
4. Modeling of Monte-Carlo does not account for the details of the distribution of radiation inside a single cell.

These reasons have defined the development of a new approach of mathematical modeling of the interaction of light with biological particles and biological tissue through the application of asymptotic methods in the theory of diffraction.

This approach enabled:

1. the investigation of the optical properties of an ensemble of randomly oriented spherical particles (hemocytes) in the cavity optical linear resonator;
2. the calculation of the refractive index of the blood and to determine the speed blood of an flow in the capillary at a wavelength $\lambda = 0.63\,\mu$m for the case in vivo;

3. the investigation of the optical characteristics of the simulated biological structure with roughness, when the characteristic size of unevenness on the surface is much greater than the wavelength, by the classical methods of the theory of diffraction;
4. the evaluation of the effect of roughness on the spectral characteristics of the simulated biological structure;
5. the calculation of the preliminary parameters of the laser radiation field, to identify and study the effects of responses of laser irradiation at different levels of organization of living matter;
6. the description of the quantitatively and qualitatively normalized spectra of laser radiation on the oxy-and deoxygemoglobin and the selection of the optimal wavelength for the effective action of laser radiation on biological structures;
7. the study the effectiveness of absorption not only by blood but also in biological tissues, and the investigation of the kinetics of the denaturation of tissue in order to develop the optimal mode of operation and technical characteristics of laser used in biomedical research;
8. theoretically calculate the size distribution function for particles of irregular shape with a variety forms and structures of inclusions that simulate blood cells in the case of in vivo and determine the degree of aggregation, for example, the platelet for case in vivo.

References

1. G. Muller et al. (eds.), *Medical Optical Tomography: Functional Imaging and Monitoring* (Bellinhgham, SPIE, 1993) IS11
2. G.R. Ivanitskii, A.S. Kunisky, *Study of the Microstructure Objects by Means of Coherent Optics* (Moscow, 1981)
3. V.V. Lopatin, F.Ya Sidko, *The Polarization Characteristics of Suspensions of Biological Particles* (Novosibirsk, 1991)
4. A. Brunsting, P.F. Mullaney, Differential light scattering from spherical mammalian cells. Biophys. J. **14**(N6), 439–453 (1974)
5. P.F. Mullaney, R.J. Fiel, Cellular stucture as revealed by visible light scattering: studies on suspensions of red blood cell ghost. Appl. Opt. **15**(2), 301–311 (1976)
6. A. Brunsting, P.F. Mullaney, Light scattering from coated spheres: model for biological cells. Appl. Opt. **11**(3), 675–680 (1972)
7. A. Brunsting, P.F. Mullaney, Differential light scattering: possible method of mammalian cell indentification. J. Colloid Interface Sci. **39**(3), 492–496 (1972)
8. P. Latimer, Light scattering by homogeneous sphere with radial projections. Appl. Opt. **23**(3), 442–447 (1984)
9. P. Latimer, Light scattering, data inversion, and information theory. J. Colloid Interface Sci. **39**(3), 497–503 (1972)
10. P. Latimer, Light scattering and absorpition as method of studying cell population parameters. Ann. Rev. Biophys. Bioeng. **11**(1), 129–150 (1982)
11. P. Latimer, D.M. Moore, F.D. Bryant, Changes in total light scattering and absorpition caused by changes in particle conformation. J. Theor. Biol. **21**(N2), 348–367 (1968)
12. A. Ishimaru, *Wave Propagation and Scattering in Random Media: Single Scattering and Transport Theory*, vol. 1 (Moscow, 1981)

13. F.A. Duck, *Physical Properties of Tissue* (Academic Press, San-Diego, 1990)
14. M.H. Niemz, *Laser - Tissue Interactions: Fundamentals and Applications* (Berlin, 1996)
15. W.M. Star, in *Optical-Thermal Response of Laser-Irradiated Tissue*, ed. by A.J. Welch, M.J.C. van Gemert. Diffusion theory of light transport (Plenum, New York, 1995), pp. 131–206
16. J.B. Fishkin, E. Gratton, Propagation of photon-density waves in strongly scattering media containing an absorbing semi-infinite plane bounded by a straight edge. J. Opt. Soc. Am. A. **10**(1), 127–140 (1993)
17. V.V. Tuchin, *Tissue Optics: Light Scattering Methods and Instruments for Medical Diagnosis* (Bellingham, SPIE Tutorial Texts in Optical Enginnring (SPIE, 2000)
18. S.A. Tereshchenko, *Methods of Computer Tomography* (Moscow, 2004)
19. A.H. Hielscher, R.E. Alcouffe, Non-diffusive photon migration in homogeneous and heteregenous tissues. SPIE Proc. **2925**, 22–30 (1996)
20. G. Yoon, A.J. Welch, M. Motamedi, M.G. Van Gemert, Development and application of three dimensional light. IEEE J. Quantum Electron **23**(10), 1721–1733 (1987)
21. I.N. Minin, *The Theory of Radiative Transfer in the Atmosphere of Planets* (Moscow, 1988)
22. H. Rinneeberg, *The Inverse Problem* (Akademic Verlag, Berlin, 1995)
23. A. Ishimaru, Theory and application of wave propagation and scattering in random media. Proc. IEEE **65**(7), 1030–1060 (1977)
24. M.U. Vera, D.J. Durian, Angular distribution of diffusely transmitted light. Phys. Rev. E. **53**, 3215 (1996)
25. R.C. Haskell et al., Boundary conditions for the diffusion equation in radiative transfer. J. Opt. Soc. Am. A. **11**, 2727–2741 (1994)
26. M. Bassani et al., Independence of the diffusion coefficient from absorption: experimental and numerical evidence. Opt. Lett. **22**(N12), 853–855 (1997)
27. Van Germert M.J.C., Nelson J.S., Milner T.E. et.al Non-invasive determination of port wine stain anatomy and physiology for optimal laser treatment strategies Phys.Med.Biol.1997, Vol.42. pp.937-949
28. S.A. Prahl, M.J.C. Gemert, A.J. Welch, Determining the optical properties of turbid media by using the adding-double method. Appl. Opt. **32**, 559–568 (1993)
29. Van de Hulst H. C. Multiple Light Scattering. N.Y., 1980

Chapter 3
Study of Electrophysical Characteristics of Blood Formed Elements Using Intracavity Laser Spectroscopy

Abstract We study optical characteristics of an ensemble of arbitrarily oriented particles are studied in an optical cavity. The study is based on the self-consistent conjugation with respect to nonuniform optical cavities with the results of scattering by an ensemble of arbitrarily oriented spherical particles with different shapes and structures. A new electrodynamic model for the interaction of laser radiation with blood cells is constructed with allowance for the structure of cells for the prediction of optical properties in vivo and for study the polarization characteristics, absorption curves for an ensemble of spherical particles with a nonconcentric inclusion that are placed in a resonator cavity. Quantitative estimates are made that can be used to predict how biophysical and biochemical processes in a biological tissue may influence its optical properties.

3.1 Introduction

There has been considerable recent interest in the application of laser methods in various branches of science and technology including physics, chemistry, biology, and medicine. Laser sources are employed in medicine for diagnostics, therapy, and surgery. Informative parameters that characterize vital activity are primarily chosen in such problems. Note the important analysis of peripheral blood, which flows in organs and tissues, since the corresponding results can be used to characterize a living organism. A comprehensive analysis of the parameters of light scattering and absorption allows rapid intact determination of physiological and morphological modifications in cells due to temperature, chemical, etc., effects. It is known that blood consists of the following blood cells: leucocytes, erythrocytes, and thrombocytes [1, 2]. The study of the optical properties of such biological objects makes it possible to solve several important problems in diagnostics of pathologies. To develop a mathematical model of the interaction of laser radiation with complicated blood cells, we must consider the corresponding geometrical structures.

© Springer International Publishing AG, part of Springer Nature 2018
K. Kulikov and T. Koshlan, *Laser Interaction with Heterogeneous
Biological Tissue*, Biological and Medical Physics, Biomedical Engineering,
https://doi.org/10.1007/978-3-319-94114-1_3

First, we consider the cells with the highest concentration in blood (erythrocytes). An erythrocyte is a cell that has the shape of a biconcave disk. The cell does not contain a nucleus and a specific protein (hemoglobin) is the main component of cytoplasm. In normal blood, from 70 to 80% of erythrocytes have the spherical biconcave shape and different shapes are possible for the remaining 20–30% of the cells (e.g., spherical, oval, bowl-shaped, etc.). The erythrocyte shape is sensitive to several diseases: in particular, sickle cells are typical of sicklemia [1, 2].

Leucocytes are blood cells that can be divided into granulocytes, which exhibit granules, and agranulocytes, which are free of granules. Neutrophils, eosinophils, and basophils are classified as granulocytes [1, 2].

A neutrophil is a circular cell with an uncommon rod-shaped nucleus. Neutrophils with rod-shaped and lobed nuclei are young and mature cells, respectively. Most neutrophils in blood are cells with lobed nuclei (65%), and the content of the plane-nucleus cells is no greater than 5%.

Similarly to the neutrophil, an eosinophil is a circular cell with rod-shaped or lobed nucleus. The cytoplasm of this cell contains relatively large granules with identical sizes and shapes [1, 2].

A basophil is a circular cell with the rod-shaped or lobed nucleus. The cytoplasm contains granules with different sizes and shapes [1, 2].

Monocytes and lymphocytes are classified as agranulocytes.

Monocytes and lymphocytes are classified as agranulocytes. A monocyte is an agranulocyte (i.e., a cell that does not contain granules) with an almost triangular shape and a large nucleus that can be circular, beanlike, etc.

A lymphocyte is a circular cell with a variable size and a relatively large circular nucleus. Lymphocytes are formed from lymphoblasts in bone marrow, where the remaining blood cells are formed, and exhibit several divisions in the course of maturation.

A thrombocyte is a relatively small circular or oval nucleus-free cell. In this work, we construct an electrodynamic model of the interaction of laser radiation with blood cells for the prediction of the electrophysical properties. Optical intracavity methods are efficient tools for the study of processes in complicated biological systems.

The problem consists of three consecutive stages.

At the first stage, we consider the scattering by a particle in which the nucleus is shifted relative to the center. Note the variable position of the nucleus in the cell. In particular, the nucleus is often located at the center in young and embryo cells. The growth of the cell and an increase in the rate of metabolic processes may lead to a shift of the nucleus, which is always embedded in cytoplasm.

At the second stage, we solve the problem of multiple scattering by an ensemble of spheres that is used to simulate the biological medium (blood formed elements) in the optical cavity. In this case, we self-consistently take into account the multiple scattering by a set of particles with nonconcentric inclusions and propose a solution for the eigenfrequencies of the optical cavity with a cuvette that contains particles with complicated structures.

At the third stage, we investigate the numerical problem of optical characteristics for an ensemble of spherical particles with a non concentric inclusion that are placed in a resonator cavity.

Chapter is based on the results of the [3, 4].

3.2 Vector Spherical Harmonics

We assume that a cuvette with a sample of biotissues that simulates blood formed elements is placed in the vicinity of the Z axis in region Ω of a linear cavity. In the first approximation, we also assume that the particles that simulate blood formed elements, in particular, erythrocytes are spherical particles and the remaining blood formed elements are represented as spheres with nonconcentric inclusions (see Fig. 3.1).

We assume that the particle sizes exceed the incident radiation wavelength; i.e., $ka^j > 1$, where a^j is the radius of the jth particle.

Let a plane, linearly polarized electromagnetic wave be incident on a group of uniform particles with radii a^j and refractive indices $N^j = n^{(o)j} + i\chi^j$, where j is the particle number. The wave propagates in a random direction. The particle ensemble is considered in a three-dimensional coordinate system with the origin at the center of the particle j_0. The radius vector of any other jth particle is denoted as $\mathbf{r}_{j_0, j}$. The field in the vicinity of the j_0-particle, perturbed by other particles, is determined from the Maxwell equations

$$\mathrm{rot}\mathbf{H} = ik\mathbf{E}, \quad \mathrm{rot}\mathbf{E} = -ik\,\mathbf{H}, \quad \mathrm{div}\mathbf{E} = 0, \quad \mathrm{div}\mathbf{H} = 0,$$

where k is the wave number.

Let us introduce a vector such that $\mathbf{M} = \nabla \times (\mathbf{r}\psi)$, where ψ is a scalar function and $\mathbf{r}$ is the radius vector, $\nabla \cdot \mathbf{M} = 0$. If we use the vector identities,

$$\nabla \times (\mathbf{A} \times \mathbf{B}) = \mathbf{A}(\nabla \cdot \mathbf{B}) - \mathbf{B}(\nabla \cdot \mathbf{A}) + (\mathbf{B} \cdot \nabla)\mathbf{A} - (\mathbf{A} \cdot \nabla)\mathbf{B},$$

$$\nabla \cdot (\mathbf{A} \cdot \mathbf{B}) = \mathbf{A} \times (\nabla \times \mathbf{B}) + \mathbf{B} \times (\nabla \times \mathbf{A}) + \mathbf{B}(\nabla \cdot \mathbf{A}) + (\mathbf{A} \cdot \nabla)\mathbf{B},$$

Fig. 3.1 Linear resonator with the cell containing the erythrocyte monolayer

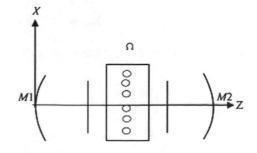

then we obtain

$$\nabla^2 \mathbf{M} + k^2 \mathbf{M} = \nabla \times (\mathbf{r}(\nabla^2 \psi + k^2 \psi)). \tag{3.1}$$

From (3.1) it follows that $\mathbf{M}$ satisfies the wave equation if ψ is a solution to the scalar wave equation

$$\nabla^2 \psi + k^2 \psi = 0 \tag{3.2}$$

Then

$$\mathbf{M} = -\mathbf{r} \times \nabla \psi$$

whence it follows that $\mathbf{M}$ is perpendicular to $\mathbf{r}$.

Let us construct from $\mathbf{M}$ another vector function

$$\mathbf{N} = \frac{1}{k} \nabla \times \mathbf{M},$$

which also satisfies the vector wave equation

$$\nabla^2 \mathbf{N} + k^2 \mathbf{N} = 0.$$

Therefore, $\mathbf{M}$ and $\mathbf{N}$ have all the required properties of an electromagnetic field: they satisfy the vector wave equation, they are divergence-free, the curl of $\mathbf{M}$ is proportional to $\mathbf{N}$, and the curl of $\mathbf{N}$ is proportional to $\mathbf{M}$. Thus, the problem of finding solutions to the field equations reduces to the comparatively simpler problem of finding solutions to the scalar wave equation. We shall call the scalar function ψ a generating function for the vector harmonics $\mathbf{M}$ and $\mathbf{N}$; the vector $\mathbf{r}$ is sometimes called the guiding vector. The choice of generating functions is dictated by whatever symmetry may exist in the problem. In this chapter we are interested in scattering by a sphere; therefore, we choose functions ψ that satisfy the wave equation in spherical polar coordinates. Let us rewrite scalar wave equation (3.2) in the spherical coordinate system.

$$\frac{1}{r^2}\frac{\partial}{\partial r}\left(r^2\frac{\partial \psi}{\partial r}\right) + \frac{1}{r^2 \sin\theta}\frac{\partial}{\partial\theta}\left(\sin\theta\frac{\partial \psi}{\partial\theta}\right) + \frac{1}{r^2 \sin^2\theta}\frac{\partial^2 \psi}{\partial\phi^2} + k^2\psi = 0$$

The solution of this equation in the spherical coordinate system has the form:

$$\psi_{mn} = P_n^m(\cos\theta)e^{im\phi}z_n^J(kr),$$

where z_n^J is any of the four spherical functions:

$$j_n(p) = \sqrt{\frac{\pi}{2p}}\, J_{n+\frac{1}{2}}(p), \quad y_n(p) = \sqrt{\frac{\pi}{2p}}Y_{n+\frac{1}{2}}(p), \tag{3.3}$$

$$h_n^{(1)} = j_n(p) + i,\ y_n(p), \quad h_n^{(2)} = j_n(p) - iy_n(p). \tag{3.4}$$

The vector spherical harmonics produced by ψ_{mn} are:

$$\mathbf{M}_{mn} = \nabla \times (\mathbf{r}\psi_{mn}), \quad \mathbf{N}_{mn} = \frac{\nabla \times \mathbf{M}_{mn}}{k}. \tag{3.5}$$

Writing expressions (3.5) component by component, we obtain:

$$\mathbf{M}_{mn}^J = \left[\frac{m}{\sin\theta} P_n^m(\cos\theta)i\,\mathbf{e}_\theta - \frac{\partial}{\partial\theta} P_n^m(\cos\theta)\mathbf{e}_\phi\right] z_n^J(kr)e^{im\phi}, \tag{3.6}$$

$$\mathbf{N}_{mn}^J = n(n+1)P_n^m(\cos\theta)\mathbf{e}_\mathbf{r}\frac{z_n^J(kr)}{kr}e^{im\phi} + \frac{\partial}{\partial\theta}P_n^m(\cos\theta)\frac{1}{kr}\frac{\partial}{\partial}rz_n^J(kr)\times$$

$$\times e^{im\phi}\mathbf{e}_\theta + i\frac{m}{\sin\theta}P_n^m(\cos\theta)\frac{1}{kr}\frac{\partial}{\partial r}\left[rz_n^J(kr)e^{im\phi}\right]\mathbf{e}_\phi \tag{3.7}$$

The vector harmonics $\mathbf{M}_{mn}^J$, $\mathbf{N}_{mn}^J$ will be used when solving the problem of scattering at a random jth particle surrounded by other scattering particles of arbitrary radii and refractive indices. When found, this solution will be used as a constituent to solve the more complicated problem of the epigenous of the optical cavity with an ensemble of scattering particles inside.

3.3 Scattering by a Particle with a Shifted Nucleus

Note the practical interest in the solution to the problem of scattering by a particle in which the nucleus is shifted relative to the center, since the central position of the nucleus is analyzed in [5, 6].

In this section, we consider the scattering by biological particles, in particular, blood formed elements with spherical shapes and complicated structures, since the presence of the nucleus and cytoplasm is possible. We neglect the cellular membrane, since it is very thin and insignificantly affects the light scattering. Figure 3.2 demonstrates the scattering geometry. Here, a is the radius of cell nucleus and b is the radius of cytoplasm. Figure 3.2 demonstrates the scattering geometry. Here, a is the radius of cell nucleus and b is the radius of cytoplasm.

We expand the wave that is incident on the surface of the jth particle in terms of vector spherical harmonics. Thus, we obtain

$$\mathbf{E}_i(j) = \sum_{n=1}^{\infty}\sum_{m=-n}^{n} E_{nm}[p_{nm}^j\mathbf{N}_{nm}^1 + q_{nm}^j\mathbf{M}_{nm}^1], \tag{3.8}$$

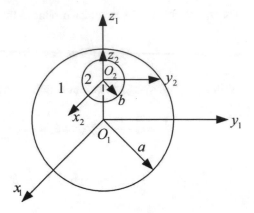

Fig. 3.2 Scattering by a spherical particle with a nonconcentric inclusion

$$\mathbf{H_i}(j) = \frac{k}{\omega\mu} \sum_{n=1}^{\infty} \sum_{m=-n}^{n} E_{nm}[q_{nm}^j \mathbf{N}_{nm}^1 + p_{nm}^j \mathbf{M}_{nm}^1]. \tag{3.9}$$

Consider the expression for the field that is scattered by the jth particle. At relatively large distances from the particle, the scattered field must be a divergent spherical wave. Therefore, we employ functions $h_n^{(1)}$:

$$h_n^{(1)} \sim (-i)^n \exp[ikr]/[ikr], \quad kr \gg n^2$$

Then, we have

$$\mathbf{E}_s(j) = \sum_{n=1}^{\infty} \sum_{m=-n}^{n} E_{nm}[a_{nm}^j \mathbf{N}_{nm}^3 + b_{nm}^j \mathbf{M}_{nm}^3],$$

$$\mathbf{H}_s(j) = \frac{k}{\omega\mu} \sum_{n=1}^{\infty} \sum_{m=-n}^{n} E_{nm}[b_{nm}^j \mathbf{N}_{nm}^3 + a_{nm}^j \mathbf{M}_{nm}^3].$$

The field in the vicinity of the center of the sphere of the jth particle is represented as

$$\mathbf{E}_{I_1}(j) = \sum_{n=1}^{\infty} \sum_{m=-n}^{n} i E_{nm}[d_{nm_1}^j \mathbf{N}_{nm}^1 + c_{nm_1}^j \mathbf{M}_{nm}^1], \tag{3.10}$$

$$\mathbf{H}_{I_1}(j) = \frac{k_1^j}{\omega\mu_1^j} \sum_{n=1}^{\infty} \sum_{m=-n}^{n} E_{nm}[c_{nm_1}^j \mathbf{N}_{nm}^1 + d_{nm_1}^j \mathbf{M}_{nm}^1]. \tag{3.11}$$

For the jth particle, the field in interval $b \leq r \leq a$ (in the $O_1 x_1 y_1 z_1$ coordinate system) is written as

$$\mathbf{E}_{I_2}(j) = \sum_{n=1}^{\infty} \sum_{m=-n}^{n} E_{nm} [d_{nm_{2_o}}^j \mathbf{N}_{nm}^3 + c_{nm_{2_o}}^j \mathbf{M}_{nm}^3 +$$

$$+ f_{nm_{2_o}}^j \mathbf{M}_{nm}^4 + g_{nm_{2_o}}^j \mathbf{N}_{nm}^4], \tag{3.12}$$

$$\mathbf{H}_{I_2}(j) = \frac{k_2^j}{\omega \mu_2^j} \sum_{n=1}^{\infty} \sum_{m=-n}^{n} E_{nm} [d_{nm_{2_o}}^j \mathbf{M}_{nm}^3 + c_{nm_{2_o}}^j \mathbf{N}_{nm}^3 +$$

$$+ f_{nm_{2_o}}^j \mathbf{N}_{nm}^4 + g_{nm_{2_o}}^j \mathbf{M}_{nm}^4]. \tag{3.13}$$

The field of the spherical inclusion of the jth particle (in the $O_2 x_2 y_2 z_2$ coordinate system) is represented as

$$\mathbf{E}_{I_{2\,(\text{inclusion})}}(j) = \sum_{n=1}^{\infty} \sum_{m=-n}^{n} E_{nm} [d_{nm_2}^j \mathbf{N}_{nm}^3 + c_{nm_2}^j \mathbf{M}_{nm}^3 +$$

$$+ f_{nm_2}^j \mathbf{M}_{nm}^4 + g_{nm_2}^j \mathbf{N}_{nm}^4], \tag{3.14}$$

$$\mathbf{H}_{I_{2\,(\text{inclusion})}}(j) = \frac{k_2^j}{\omega \mu_2^j} \sum_{n=1}^{\infty} \sum_{m=-n}^{n} E_{nm} [d_{nm_2}^j \mathbf{M}_{nm}^3 + c_{nm_2}^j \mathbf{N}_{nm}^3 +$$

$$+ f_{nm_2}^j \mathbf{N}_{nm}^4 + g_{nm_2}^j \mathbf{M}_{nm}^4], \tag{3.15}$$

where

$$E_{mn} = |E_0| i^n [2n+1] \frac{(n-m)!}{(n+m)!}.$$

To determine scattering coefficients a_{mn}^j and b_{mn}^j for the spherical particle with the shifted nucleus, we must use summation theorems based on the recurrence approach in the calculation of scalar and vector coefficients that emerge due to translation of spherical vector harmonics from the coordinate system centered at the main sphere to the coordinate system that is bound to the center of the spherical inclusion [7]:

$$\mathbf{M}_{nm,2}^{(q)} = \sum_{n'=0}^{\infty} A_{n'n}^{m,q} \mathbf{M}_{n'm,1}^{(q)} + B_{n'n}^{m,q} \mathbf{N}_{n'm,1}^{(q)}, \tag{3.16}$$

$$\mathbf{N}_{nm,2}^{(q)} = \sum_{n'=0}^{\infty} B_{n'n}^{m,q} \mathbf{M}_{n'm,1}^{(q)} + A_{n'n}^{m,q} \mathbf{N}_{n'm,1}^{(q)}, \tag{3.17}$$

Here, q is the order of the spherical Bessel functions ($q = 3, 4$). This relationship is valid at $r > |d|$, where d is the intercenter distance and $A_{n'}^{n,m,q}$ and $B_{n'}^{n,m,q}$ are given by [7, 8]

$$
A_{n'}^{n,m,q} = C_{n'}^{(n,m,q)} - \frac{k_1 d}{n'+1} \sqrt{\frac{(n'-m+1)(n'+m+1)}{(2n'+1)(2n'+3)}} C_{n'+1}^{(n,m,q)} -
$$

$$
\frac{k_1 d}{n'} \sqrt{\frac{(n'-m)(n'+m)}{(2n'+1)(2n'-1)}} C_{n'-1}^{(n,m,q)},
\tag{3.18}
$$

$$
B_{n'}^{n,m,q} = \frac{-ik_1 m d}{n'(n'+1)} C_{n'}^{(nm,q)}, \quad C_{n'}^{(0,0,q)} = \sqrt{(2n'+1)} j_n'(k_1 d),
$$

$$
C_{n'}^{(-1,0,q)} = -\sqrt{(2n'+1)} j_n'(k_1 d),
\tag{3.19}
$$

$$
C_{n'}^{(n+1,0,q)} = \frac{1}{n+1} \sqrt{\frac{2n+3}{2n'+1}} \left[n' \sqrt{\frac{2n+1}{2n'-1}} C_{n'-1}^{(n,0,q)} + n \sqrt{\frac{2n'+1}{2n-1}} C_{n'}^{(n-1,0,q)} \right]
$$

$$
- \frac{1}{n+1} \sqrt{\frac{2n+3}{2n'+1}} \left[(n'+1) \sqrt{\frac{2n+1}{2n'+3}} C_{n'+1}^{(n,0,q)} \right],
\tag{3.20}
$$

$$
C_{n'}^{(n,m,q)} = \frac{\sqrt{(n'-m+1)(n'+m)(2n'+1)}}{\sqrt{(n-m+1)(n+m)(2n'+1)}} C_{n'}^{(n,m-1,q)} -
$$

$$
-k_1 d \sqrt{\frac{(n'-m+2)(n'-m+1)}{(2n'+3)(n-m+1)(n+m)(2n'+1)}} C_{n'+1}^{(n,m-1,q)} -
$$

$$
-k_1 d \sqrt{\frac{(n'+m)(n'+m-1)}{(2n'-1)(n-m+1)(n+m)(2n'+1)}} C_{n'-1}^{(n,m-1,q)},
$$

$$
C_{n'}^{(n,m,q)} = C_{n'}^{(n,-m,q)},
\tag{3.21}
$$

$$
A_{n'}^{(n,m,3)} = A_{n'}^{(n,m,4)} = A_{n'}^{(n,-m,3)} = A_{n'}^{(n,m)},
\tag{3.22}
$$

$$
B_{n'}^{(n,m,3)} = B_{n'}^{(n,m,4)} = B_{n'}^{(n,-m,3)} = B_{n'}^{(n,m)},
\tag{3.23}
$$

$$
C_{n'}^{(n,m,3)} = C_{n'}^{(n,m,4)} = C_{n'}^{(n,-m,3)} = C_{n'}^{(n,m)}.
\tag{3.24}
$$

When $d = 0$, we have $A_{n'}^{(n,m)} = \delta_{n'n}$, $B_{n'}^{(n,m)} = 0$. We substitute expressions (3.16) and (3.17) in formulas (3.14) and (3.15) to obtain

$$\mathbf{E}_{I_{2\,(\text{inclusion})}}(j) = \sum_{n=1}^{\infty} \sum_{m=-n}^{n} E_{nm} \left[d_{nm_2}^{j} \left[\sum_{n'=0}^{\infty} A_{n'n}^{m,3} \mathbf{M}_{n'm,1}^{(3)} + B_{n'n}^{m,3} \mathbf{N}_{n'm,1}^{(3)} \right] \right] +$$

$$+ E_{nm} \left[c_{nm_2}^{j} \left[\sum_{n'=0}^{\infty} B_{n'n}^{m,3} \mathbf{M}_{n'm,1}^{(3)} + A_{n'n}^{m,3} \mathbf{N}_{n'm,1}^{(3)} \right] \right] +$$

$$+ E_{nm} \left[f_{nm_2}^{j} \left[\sum_{n'=0}^{\infty} A_{n'n}^{m,4} \mathbf{M}_{n'm,1}^{(4)} + B_{n'n}^{m,4} \mathbf{N}_{n'm,1}^{(4)} \right] \right] +$$

$$E_{nm} \left[g_{nm_2}^{j} \left[\sum_{n'=0}^{\infty} B_{n'n}^{n,m,4} \mathbf{M}_{n'm,1}^{(3)} + A_{n'n}^{m,4} \mathbf{N}_{n'm,1}^{(4)} \right] \right], \qquad (3.25)$$

$$\mathbf{H}_{I_{2\,(\text{inclusion})}}(j) = \frac{k_2^{j}}{\omega \mu_2^{j}} \sum_{n=1}^{\infty} \sum_{m=-n}^{n} E_{nm} \left[c_{nm_2}^{j} \left[\sum_{n'=0}^{\infty} A_{n'n}^{m,3} \mathbf{M}_{n'm,1}^{(3)} + B_{n'n}^{m,3} \mathbf{N}_{n'm,1}^{(3)} \right] \right] +$$

$$+ E_{nm} \left[d_{nm_2}^{j} \left[\sum_{n'=0}^{\infty} B_{n'n}^{m,3} \mathbf{M}_{n'm,1}^{(3)} + A_{n'n}^{m,3} \mathbf{N}_{n'm,1}^{(3)} \right] \right] +$$

$$+ E_{nm} \left[g_{nm_2}^{j} \left[\sum_{n'-0}^{\infty} A_{n'n}^{m,4} \mathbf{M}_{n'm,1}^{(4)} + B_{n'n}^{m,4} \mathbf{N}_{n'm,1}^{(4)} \right] \right] +$$

$$E_{nm} \left[f_{nm_2}^{j} \left[\sum_{n'=0}^{\infty} B_{n'n}^{n,m,4} \mathbf{M}_{n'm,1}^{(3)} + A_{n'n}^{m,4} \mathbf{N}_{n'm,1}^{(4)} \right] \right], \qquad (3.26)$$

These expressions are represented as

$$\mathbf{E}_{I_{2\,(\text{inclusion})}}(j) = \sum_{n=1}^{\infty} \sum_{m=-n}^{n} \sum_{n'=0}^{\infty} E_{nm} \left[d_{nm_2}^{j} A_{n'n}^{m,3} + c_{nm_2}^{j} B_{n'n}^{m,3} \right] \mathbf{M}_{n'm,1}^{(3)} +$$

$$+ \sum_{n=1}^{\infty} \sum_{m=-n}^{n} \sum_{n'=0}^{\infty} E_{nm} \left[c_{nm_2}^{j} A_{n'n}^{m,3} + d_{nm_2}^{j} B_{n'n}^{m,3} \right] \mathbf{N}_{n'm,1}^{(3)} +$$

$$+ \sum_{n=1}^{\infty} \sum_{m=-n}^{n} \sum_{n'=0}^{\infty} E_{nm} \left[f_{nm_2}^{j} A_{n'n}^{m,4} + g_{nm_2}^{j} B_{n'n}^{m,4} \right] \mathbf{M}_{n'm,1}^{(4)} + \qquad (3.27)$$

$$+ \sum_{n=1}^{\infty} \sum_{m=-n}^{n} \sum_{n'=0}^{\infty} E_{nm} \left[g_{nm_2}^{j} A_{n'n}^{m,4} + + f_{nm_2}^{j} B_{n'n}^{m,4} \right] \mathbf{N}_{n'm,1}^{(4)},$$

$$\mathbf{H}_{I_{2\,(\text{inclusion})}}(j) = \frac{k_2^j}{\omega\mu_2^j} \sum_{n=1}^{\infty} \sum_{m=-n}^{n} \sum_{n'=0}^{\infty} E_{nm} \left[d_{nm_2}^j A_{n'n}^{m,3} + c_{nm_2}^j B_{n'n}^{m,3} \right] \mathbf{N}_{n'm,1}^{(3)} +$$

$$+ \sum_{n=1}^{\infty} \sum_{m=-n}^{n} \sum_{n'=0}^{\infty} E_{nm} \left[c_{nm_2}^j A_{n'n}^{m,3} + d_{nm_2}^j B_{n'n}^{m,3} \right] \mathbf{M}_{n'm,1}^{(3)} +$$

$$+ \sum_{n=1}^{\infty} \sum_{m=-n}^{n} \sum_{n'=0}^{\infty} E_{nm} \left[f_{nm_2}^j A_{n'n}^{m,4} + g_{nm_2}^j B_{n'n}^{m,4} \right] \mathbf{N}_{n'm,1}^{(4)} + \qquad (3.28)$$

$$+ \sum_{n=1}^{\infty} \sum_{m=-n}^{n} \sum_{n'=0}^{\infty} E_{nm} \left[g_{nm_2}^j A_{n'n}^{m,4} + f_{nm_2}^j B_{n'n}^{m,4} \right] \mathbf{M}_{n'm,1}^{(4)}$$

The comparison of expressions (3.27) and (3.12) and expressions (3.28) and (3.15) yields the following relationships:

$$\sum_{n=1}^{\infty} \sum_{m=-n}^{n} c_{nm_{2o}}^j \mathbf{M}_{nm}^3 = \sum_{n=1}^{\infty} \sum_{m=-n}^{n} \sum_{n'=0}^{\infty} \left[d_{nm_2}^j A_{n'n}^{m,3} + c_{nm_2}^j B_{n'n}^{m,3} \right] \mathbf{M}_{n'm}^{(3)} \qquad (3.29)$$

$$\sum_{n=1}^{\infty} \sum_{m=-n}^{n} d_{nm_{2o}}^j \mathbf{N}_{nm}^3 = \sum_{n=1}^{\infty} \sum_{m=-n}^{n} \sum_{n'=0}^{\infty} \left[c_{nm_2}^j A_{n'n}^{m,3} + d_{nm_2}^j B_{n'n}^{m,3} \right] \mathbf{N}_{n'm,1}^{(3)} \qquad (3.30)$$

$$\sum_{n=1}^{\infty} \sum_{m=-n}^{n} f_{nm_{2o}}^j \mathbf{M}_{nm}^4 = \sum_{n=1}^{\infty} \sum_{m=-n}^{n} \sum_{n'=0}^{\infty} \left[f_{nm_2}^j A_{n'n}^{m,4} + g_{nm_2}^j B_{n'n}^{m,4} \right] \mathbf{M}_{n'm,1}^{(4)} \qquad (3.31)$$

$$\sum_{n=1}^{\infty} \sum_{m=-n}^{n} g_{nm_{2o}}^j \mathbf{N}_{nm}^4 = \sum_{n=1}^{\infty} \sum_{m=-n}^{n} \sum_{n'=0}^{\infty} \left[g_{nm_2}^j A_{n'n}^{m,4} + f_{nm_2}^j B_{n'n}^{m,4} \right] \mathbf{N}_{n'm,1}^{(4)} \qquad (3.32)$$

$$\sum_{n=1}^{\infty} \sum_{m=-n}^{n} c_{nm_{2o}}^j \mathbf{N}_{nm}^3 = \sum_{n=1}^{\infty} \sum_{m=-n}^{n} \sum_{n'=0}^{\infty} \left[d_{nm_2}^j A_{n'n}^{m,3} + c_{nm_2}^j B_{n'n}^{m,3} \right] \mathbf{N}_{n'm}^{(3)} \qquad (3.33)$$

$$\sum_{n=1}^{\infty} \sum_{m=-n}^{n} d_{nm_{2o}}^j \mathbf{M}_{nm}^3 = \sum_{n=1}^{\infty} \sum_{m=-n}^{n} \sum_{n'=0}^{\infty} \left[c_{nm_2}^j A_{n'n}^{m,3} + d_{nm_2}^j B_{n'n}^{m,3} \right] \mathbf{M}_{n'm,1}^{(3)} \qquad (3.34)$$

$$\sum_{n=1}^{\infty} \sum_{m=-n}^{n} f_{nm_{2o}}^j \mathbf{N}_{nm}^4 = \sum_{n=1}^{\infty} \sum_{m=-n}^{n} \sum_{n'=0}^{\infty} \left[f_{nm_2}^j A_{n'n}^{m,4} + g_{nm_2}^j B_{n'n}^{m,4} \right] \mathbf{N}_{n'm,1}^{(4)} \qquad (3.35)$$

$$\sum_{n=1}^{\infty} \sum_{m=-n}^{n} g_{nm_{2o}}^j \mathbf{M}_{nm}^4 = \sum_{n=1}^{\infty} \sum_{m=-n}^{n} \sum_{n'=0}^{\infty} \left[g_{nm_2}^j A_{n'n}^{m,4} + f_{nm_2}^j B_{n'n}^{m,4} \right] \mathbf{M}_{n'm,1}^{(4)} \qquad (3.36)$$

Scalar multiplication of expression (3.29) by $\mathbf{M}_{ki}^3$ and integration using orthogonality of spherical harmonics yield

$$c_{ki_{2o}}^j = \sum_{n=0}^{\infty} \left[d_{ni_2}^j A_{nk}^{i,3} + c_{ni_2}^j B_{n,k}^{i,3} \right], \tag{3.37}$$

This expression can be represented as

$$c_{nm_{2o}}^j = \sum_{n'=0}^{\infty} \left[d_{n'm_2}^j A_{n'n}^{m,3} + c_{n'm_2}^j B_{n'n}^{m,3} \right], \tag{3.38}$$

Similar relationships are obtained for remaining expressions (3.30)–(3.32):

$$c_{nm_{2o}}^j = \sum_{n'=0}^{\infty} \left[d_{n'm_2}^j A_{n'n}^{m,3} + c_{n'm_2}^j B_{n'n}^{m,3} \right], \tag{3.39}$$

$$d_{nm_{2o}}^j = \sum_{n'=0}^{\infty} \left[c_{n'm_2}^j A_{n'n}^{m,3} + d_{n'm_2}^j B_{n'n}^{m,3} \right], \tag{3.40}$$

$$f_{nm_{2o}}^j = \sum_{n'=0}^{\infty} \left[f_{n'm_2}^j A_{n'n}^{m,4} + g_{n'm_2}^j B_{n'n}^{m,4} \right], \tag{3.41}$$

$$g_{nm_{2o}}^j = \sum_{n'=0}^{\infty} \left[g_{n'm_2}^j A_{n'n}^{m,4} + f_{n'm_2}^j B_{n'n}^{m,4} \right]. \tag{3.42}$$

With allowance for formulas (3.23) and (3.24), expressions (3.40) and (3.42) are represented as

$$d_{nm_{2o}}^j = \sum_{n'=0}^{\infty} \left[d_{n'm_2}^j A_{n'n}^m + c_{n'm_2}^j B_{n'n}^m \right], \tag{3.43}$$

$$c_{nm_{2o}}^j = \sum_{n'=0}^{\infty} \left[c_{n'm_2}^j A_{n'n}^m + d_{n'm_2}^j B_{n'n}^{m,3} \right], \tag{3.44}$$

$$f_{nm_{2o}}^j = \sum_{n'=0}^{\infty} \left[f_{n'm_2}^j A_{n'n}^m + g_{n'm_2}^j B_{n'n}^{m,4} \right], \tag{3.45}$$

$$g_{nm_{2o}}^j = \sum_{n'=0}^{\infty} \left[g_{n'm_2}^j A_{n'n}^m + f_{n'm_2}^j B_{n'n}^m \right]. \tag{3.46}$$

Expressions (3.43) and (3.46) establish relationships of the coefficients for the sphere that is located at the center of the coordinate system and the sphere that is shifted relative to the center.

The boundary conditions are written as

$$[\mathbf{E}_{I_1}(j) - \mathbf{E}_{I_2}(j)]|_{r_2=b} \times \mathbf{e_r} = 0, \quad [\mathbf{H}_{I_1}(j) - \mathbf{H}_{I_2}(j)]|_{r_2=b} \times \mathbf{e_r} = 0, \qquad (3.47)$$

$$[\mathbf{E}_i(j) + \mathbf{E_s}(j) - \mathbf{E}_{I_1}(j)]|_{r_1=a} \times \mathbf{e_r} = [\mathbf{H}_i(j) + \mathbf{H}_s(j) - \mathbf{H}_{I_1}(j)]_{r_1=a} \times \mathbf{e_r} \qquad (3.48)$$

$$E_{I_{1\theta}}(j)|_{r_2=b} = E_{I_{2\theta}}(j)|_{r_2=b}, \, E_{I_{1\phi}}(j)|_{r_2=b} = E_{I_{2\phi}}(j)|_{r_2=b}, \qquad (3.49)$$

$$H_{I_{1\theta}}(j)|_{r_2=b} = H_{I_{2\theta}}(j)|_{r_2=b}, \, H_{I_{1\phi}}(j))|_{r_2=b} = H_{I_{2\phi}}(j)|_{r_2=b}, \qquad (3.50)$$

$$E_{i\theta}(j)|_{r_1=a} + E_{s\theta}(j)|_{r_1=a} = E_{I_{1\theta}}(j)|_{r_1=a}, \qquad (3.51)$$

$$E_{i\phi}(j)|_{r_1=a} + E_{s\phi}(j)|_{r_1=a} = E_{I_{1\phi}}(j)|_{r_1=a}, \qquad (3.52)$$

$$H_{i\theta}(j)|_{r_1=a} + H_{s\theta}(j)|_{r_1=a} = H_{I_{1\theta}}(j)|_{r_1=a}, \qquad (3.53)$$

$$H_{i\phi}(j)|_{r_1=a} + H_{s\phi}(j)|_{r_1=a} = H_{I_{1\phi}}(j)|_{r_1=a}. \qquad (3.54)$$

Substituting the expressions for the fields that are expanded in terms of vector spherical harmonics with allowance for relationships (3.43)−(3.46) in boundary conditions (3.49)−(3.54) and taking into account the orthogonality of the spherical harmonics, we derive a system of equations for unknown coefficients. Note that the scattering coefficients that are found using this system can be represented as

$$a_{mn}^j = a_{n_{1p}}^j p_{mn}^j + a_{n_{1q}}^j q_{mn}^j, b_{mn}^j = b_{n_{1p}}^j p_{mn}^j + b_{n_{1q}}^j q_{mn}^j, \qquad (3.55)$$

where

$$b_{n_{1p}}^j = -\frac{b_{n_{11p}}^j}{b_{n_{21}}^j}, \quad b_{n_{1q}}^j = -\frac{b_{n_{11q}}^j}{b_{n_{21}}^j},$$

$$b_{n_{11p}}^j = (A_{n'n}^m)^{(2)} \psi_n((ka)^j) \xi_n'^{(2)}((ka)^j) \xi_n^{(1)}((k_1a)^j) F_{21} \xi_n'^{(1)}((k_1a)^j) +$$

$$+ (A_{n'n}^m)^2 \psi_n((ka)^j) F_{12} \xi_n'^{(1)}((k_1a)^j) \xi_n^{(1)}((ka)^j) \xi_n'^{(2)}((k_1a)^j) +$$

$$+ (A_{n'n}^m)^2 \psi_n((ka)^j) F_{12} (\xi_n'^{(1)}((k_1a)^j))^2 \xi_n^{(1)}((ka)^j) F_{21} -$$

$$- (A_{n'n}^m)^2 \psi_n((ka)^j) \xi_n'^{(2)}((k_1a)^j) \xi_n'^{(1)}((ka)^j) F_{21} \xi_n^{(1)}((k_1a)^j) -$$

$$- (A_{n'n}^m)^2 \psi_n((ka)^j) F_{12} \xi_n'^{(1)}((k_1a)^j) \xi_n'^{(1)}((ka)^j) \xi_n^{(2)}((k_1a)^j) -$$

$$-(B_{n'n}^m)^2 \psi_n((ka)^j) F_{21} \xi_n'^{(2)}((k_1a)^j) \xi_n^{(1)}((ka)^j) \xi_n'^{(1)}((k_1a)^j) -$$

$$-(B_{n'n}^m)^2 \psi_n((ka)^j) F_{12} \xi_n'^{(1)}((k_1a)^j) \xi_n^{(1)}((ka)^j) \xi_n'^{(2)}((k_1a)^j) -$$

$$-(B_{n'n}^m)^2 \psi_n((ka)^j) F_{12} F_{21} (\xi_n'^{(1)}((k_1a)^j))^2 \xi_n^{(1)}((ka)^j) +$$

$$+(B_{n'n}^m)^2 \psi_n((ka)^j) F_{21} \xi_n^{(2)}((k_1a)^j) \xi_n^{(1)}((ka)^j) \xi_n'^{(1)}((k_1a)^j) +$$

$$+(B_{n'n}^m)^2 \psi_n((ka)^j) F_{12} \xi_n^{(1)}((k_1a)^j) \xi_n^{(1)}((ka)^j) \xi_n'^{(2)}((k_1a)^j) -$$

$$-(A_{n'n}^m)^2 \psi_n'((ka)^j) F_{21} \xi_n^{(2)}((k_1a)^j) \xi_n^{(1)}((ka)^j) \xi_n'^{(1)}((k_1a)^j) -$$

$$-(A_{n'n}^m)^2 \psi_n'((ka)^j) F_{12} \xi_n^{(1)}((k_1a)^j) \xi_n^{(1)}((ka)^j) \xi_n'^{(2)}((k_1a)^j) +$$

$$+(A_{n'n}^m)^2 \psi_n'((ka)^j) F_{21} \xi_n^{(2)}((k_1a)^j) \xi_n'^{(1)}((ka)^j) \xi_n^{(1)}((k_1a)^j) +$$

$$+(A_{n'n}^m)^2 \psi_n'((ka)^j) F_{12} \xi_n^{(1)}((k_1a)^j) \xi_n'^{(1)}((ka)^j) \xi_n'^{(2)}((k_1a)^j) +$$

$$+(A_{n'n}^m)^2 \psi_n'((ka)^j) F_{12} F_{21} (\xi_n^{(1)}((k_1a)^j))^2 \xi_n'^{(1)}((ka)^j) +$$

$$+(B_{n'n}^m)^2 \psi_n'((ka)^j) F_{12} \xi_n^{(1)}((ka)^j) \xi_n^{(2)}((k_1a)^j) \xi_n'^{(1)}((k_1a)^j) +$$

$$+(B_{n'n}^m)^2 \psi_n'((ka)^j) F_{21} \xi_n^{(1)}((ka)^j) \xi_n^{(1)}((k_1a)^j) \xi_n'^{(2)}((k_1a)^j) -$$

$$-(B_{n'n}^m)^2 \psi_n'((ka)^j) F_{21} \xi_n^{(2)}((k_1a)^j) \xi_n'^{(1)}((ka)^j) \xi_n^{(1)}((k_1a)^j) -$$

$$-(B_{n'n}^m)^2 \psi_n'((ka)^j) F_{21} \xi_n^{(2)}((k_1a)^j) \xi_n'^{(1)}((ka)^j) \xi_n^{(1)}((k_1a)^j) -$$

$$-(B_{n'n}^m)^2 \psi_n'((ka)^j) F_{12} \xi_n^{(1)}((k_1a)^j) \xi_n'^{(1)}((ka)^j) \xi_n^{(2)}((k_1a)^j) -$$

$$-(B_{n'n}^m)^2 \psi_n'((ka)^j) F_{12} F_{21} (\xi_n^{(1)}((k_1a)^j))^2 \xi_n'^{(1)}((ka)^j) +$$

$$+\left[(A_{n'n}^m)^2 - (B_{n'n}^m)^2\right] \psi_n((ka)^j)(\xi_n^{(2)}((k_1a)^j))^2 \xi_n^{(1)}((ka)^j) +$$

$$+\left[(A_{n'n}^m)^2 - (B_{n'n}^m)^2\right] \psi_n'((ka)^j)(\xi_n^{(2)}((k_1a)^j))^2 \xi_n'^{(1)}((ka)^j) -$$

$$-(A_{n'n}^m)^2 \psi_n((ka)^j)(\xi_n'^{(2)}((k_1a)^j))^2 \xi_n'^{(1)}((ka)^j) \xi_n^{(2)}((k_1a)^j) +$$

$$+(B_{n'n}^m)^2 \psi_n((ka)^j)(\xi_n'^{(2)}((k_1a)^j))^2 \xi_n'^{(1)}((ka)^j) \xi_n^{(2)}((k_1a)^j) +$$

$$+(A_{n'n}^m)^2 \psi_n'((ka)^j)(\xi_n^{(2)}((k_1a)^j))^2 \xi_n^{(1)}((ka)^j) \xi_n'^{(2)}((k_1a)^j) +$$

$$+(B_{n'n}^m)^2\psi_n'((ka)^j)(\xi_n^{(2)}((k_1a)^j))\xi_n^{(1)}((ka)^j)(\xi_n^{'(2)}((k_1a)^j))-$$

$$-(A_{n'n}^m)^2\psi_n((ka)^j)F_{12}F_{21}(\xi_n^{'(1)}((k_1a)^j))\xi_n^{'(1)}((ka)^j)(\xi_n^{(1)}((k_1a)^j))+$$

$$(B_{n'n}^m)^2\psi_n((ka)^j)F_{12}F_{21}(\xi_n^{'(1)}((k_1a)^j))\xi_n^{'(1)}((ka)^j)(\xi_n^{(1)}((k_1a)^j))-$$

$$-(A_{n'n}^m)^2\psi_n'((ka)^j)F_{12}F_{21}(\xi_n^{(1)}((k_1a)^j))\xi_n^{(1)}((ka)^j)(\xi_n^{'(1)}((k_1a)^j))+$$

$$+(B_{n'n}^m)^2\psi_n'((ka)^j)F_{12}F_{21}(\xi_n^{(1)}((k_1a)^j))\xi_n^{(1)}((k_1a)^j)\xi_n^{(1)}((ka)^j),$$

$$b_{n_{11q}}^j = -(A_{n'n}^m)(B_{n'n}^m)F_{21}\psi_n((ka)^j)(\xi_n^{(2)}((k_1a)^j))^2\xi_n^{'(1)}((ka)^j)\xi_n^{'(1)}((k_1a)^j)-$$

$$-(A_{n'n}^m)(B_{n'n}^m)F_{12}\psi_n((ka)^j)(\xi_n^{(1)}((k_1a)^j))\xi_n^{(1)}((ka)^j)\xi_n^{'(2)}((k_1a)^j)+$$

$$+(A_{n'n}^m)(B_{n'n}^m)F_{21}\psi_n'((ka)^j)(\xi_n^{(2)}((k_1a)^j))\xi_n^{(1)}((ka)^j)\xi_n^{'(1)}((k_1a)^j)+$$

$$+(A_{n'n}^m)(B_{n'n}^m)F_{12}\psi_n'((ka)^j)(\xi_n^{(1)}((k_1a)^j))\xi_n^{(1)}((ka)^j)\xi_n^{'(2)}((k_1a)^j)-$$

$$-(A_{n'n}^m)(B_{n'n}^m)F_{12}\psi_n'((ka)^j)(\xi_n^{(1)}((ka)^j))\xi_n^{(2)}((k_1a)^j)\xi_n^{'(1)}((k_1a)^j)-$$

$$-(A_{n'n}^m)(B_{n'n}^m)F_{12}\psi_n'((ka)^j)(\xi_n^{(1)}((ka)^j))\xi_n^{(2)}((k_1a)^j)\xi_n^{'(1)}((k_1a)^j)-$$

$$-(A_{n'n}^m)(B_{n'n}^m)F_{21}\psi_n'((ka)^j)(\xi_n^{(1)}((ka)^j))\xi_n^{(1)}((k_1a)^j)\xi_n^{'(2)}((k_1a)^j)+$$

$$+(A_{n'n}^m)(B_{n'n}^m)F_{21}\psi_n((ka)^j)(\xi_n^{'(2)}((k_1a)^j))\xi_n^{(1)}((ka)^j)\xi_n^{(1)}((k_1a)^j)+$$

$$+(A_{n'n}^m)(B_{n'n}^m)F_{12}\psi_n((ka)^j)(\xi_n^{'(1)}((k_1a)^j))\xi_n^{'(1)}((ka)^j)\xi_n^{(2)}((k_1a)^j),$$

$$b_{n_{21}}^j = -\xi_n^{(1)}((ka)^j)(A_{n'n}^m)^2\xi_n^{'(1)}((ka)^j)\xi_n^{'(2)}((k_1a)^j)\xi_n^{(1)}((k_1a)^j)F_{21}-$$

$$-\xi_n^{(1)}((ka)^j)(A_{n'n}^m)^2\xi_n^{'(1)}((ka)^j)\xi_n^{'(1)}((k_1a)^j)\xi_n^{(2)}((k_1a)^j)F_{12}-$$

$$-\xi_n^{'(1)}((ka)^j)(A_{n'n}^m)^2\xi_n^{(1)}((ka)^j)\xi_n^{(2)}((k_1a)^j)\xi_n^{'(1)}((k_1a)^j)F_{21}-$$

$$-\xi_n^{'(1)}((ka)^j)(A_{n'n}^m)^2\xi_n^{(1)}((ka)^j)\xi_n^{'(2)}((k_1a)^j)\xi_n^{(1)}((k_1a)^j)F_{12}+$$

$$+\xi_n^{'(1)}((ka)^j)(A_{n'n}^m)^2\xi_n^{'(1)}((ka)^j)\xi_n^{(2)}((k_1a)^j)\xi_n^{(1)}((k_1a)^j)F_{21}+$$

$$+\xi_n^{'(1)}((ka)^j)(A_{n'n}^m)^2\xi_n^{(1)}((ka)^j)\xi_n^{'(2)}((k_1a)^j)\xi_n^{(1)}((k_1a)^j)F_{12}+$$

$$+\xi_n^{'(1)}((ka)^j)(A_{n'n}^m)^2\xi_n^{'(1)}((ka)^j)(\xi_n^{(1)}((k_1a)^j))^2F_{12}F_{21}+$$

$$+\xi_n^{(1)}((ka)^j)(B_{n'n}^m)^2\xi_n^{'(1)}((ka)^j)\xi_n^{'(2)}((k_1a)^j)\xi_n^{(1)}((k_1a)^j)F_{21}+$$

$$+\xi_n^{(1)}((ka)^j)(B_{n'n}^m)^2\xi_n^{'(1)}((ka)^j)\xi_n^{(2)}((k_1a)^j)\xi_n^{'(1)}((k_1a)^j)F_{12}-$$

$$-\xi_n^{'(1)}((ka)^j)(B_{n'n}^m)^2\xi_n^{'(1)}((ka)^j)\xi_n^{(2)}((k_1a)^j)\xi_n^{(1)}((k_1a)^j)F_{21}-$$

$$-\xi_n^{'(1)}((ka)^j)(B_{n'n}^m)^2\xi_n^{'(1)}((ka)^j)\xi_n^{(2)}((k_1a)^j)\xi_n^{(1)}((k_1a)^j)F_{12}-$$

$$-\xi_n^{'(1)}((ka)^j)(B_{n'n}^m)^2\xi_n^{'(1)}((ka)^j)(\xi_n^{(1)}((k_1a)^j))^2F_{21}F_{12}+$$

$$+\xi_n^{'(1)}((ka)^j)(B_{n'n}^m)^2\xi_n^{(1)}((ka)^j)\xi_n^{(2)}((k_1a)^j)\xi_n^{'(1)}((k_1a)^j)F_{21}+$$

$$+\xi_n^{'(1)}((ka)^j)(B_{n'n}^m)^2\xi_n^{(1)}((ka)^j)\xi_n^{'(2)}((k_1a)^j)\xi_n^{(1)}((k_1a)^j)F_{12}-$$

$$-\xi_n^{(1)}((ka)^j)(A_{n'n}^m)^2\xi_n^{'(1)}((ka)^j)\xi_n^{'(1)}((k_1a)^j)\xi_n^{(1)}((k_1a)^j)F_{21}F_{12}-$$

$$-\xi_n^{'(1)}((ka)^j)(A_{n'n}^m)^2\xi_n^{(1)}((ka)^j)\xi_n^{'(1)}((k_1a)^j)\xi_n^{(1)}((k_1a)^j)F_{21}F_{12}+$$

$$+\xi_n^{(1)}((ka)^j)(B_{n'n}^m)^2\xi_n^{'(1)}((ka)^j)\xi_n^{'(1)}((k_1a)^j)\xi_n^{(1)}((k_1a)^j)F_{21}F_{12}+$$

$$+\xi_n^{'(1)}((ka)^j)(B_{n'n}^m)^2\xi_n^{(1)}((ka)^j)\xi_n^{'(1)}((k_1a)^j)\xi_n^{(1)}((k_1a)^j)F_{21}F_{12}+$$

$$+\left[(A_{n'n}^m)^2-(B_{n'n}^m)^2\right](\xi_n^{'(1)}((ka)^j))^2(\xi_n^{(2)}((k_1a)^j))^2$$

$$+\left[(A_{n'n}^m)^2-(B_{n'n}^m)^2\right](\xi_n^{(1)}((ka)^j))^2(\xi_n^{'(2)}((k_1a)^j))^2+$$

$$+(\xi_n^{(1)}((ka)^j))^2(A_{n'n}^m)^2\xi_n^{'(2)}((k_1a)^j)\xi_n^{'(1)}((k_1a)^j)F_{21}+$$

$$+(\xi_n^{(1)}((ka)^j))^2(A_{n'n}^m)^2\xi_n^{'(1)}((k_1a)^j)\xi_n^{'(2)}((k_1a)^j)F_{12}+$$

$$(\xi_n^{(1)}((ka)^j))^2(A_{n'n}^m)^2(\xi_n^{'(1)}((k_1a)^j))^2F_{21}F_{12}-$$

$$-\xi_n^{(1)}((ka)^j)(A_{n'n}^m)^2\xi_n^{'(1)}((ka)^j)\xi_n^{'(2)}((k_1a)^j)\xi_n^{(2)}((k_1a)^j)-$$

$$-\xi_n^{'(1)}((ka)^j)(A_{n'n}^m)^2\xi_n^{(1)}((ka)^j)\xi_n^{'(2)}((k_1a)^j)\xi_n^{(2)}((k_1a)^j)+$$

$$+\xi_n^{(1)}((ka)^j)(B_{n'n}^m)^2\xi_n^{'(1)}((ka)^j)\xi_n^{'(2)}((k_1a)^j)\xi_n^{(2)}((k_1a)^j)-$$

$$-(\xi_n^{(1)}((ka)^j))^2(b_{n'n}^m)^2\xi_n^{'(2)}((ka)^j)\xi_n^{'(1)}((k_1a)^j)F_{21}-$$

$$-(\xi_n^{(1)}((ka)^j))^2(B_{n'n}^m)^2\xi_n^{'(1)}((k_1a)^j)\xi_n^{'(2)}((k_1a)^j)F_{12}-$$

$$(\xi_n^{(1)}((ka)^j))^2 (A_{n'n}^m)^2 (\xi_n^{'(1)}((ka)^j))^2 F_{12} F_{21} +$$

$$+\xi_n^{'(1)}((ka)^j)(B_{n'n}^m)^2 \xi_n^{(1)}((ka)^j)\xi_n^{'(2)}((k_1a)^j)\xi_n^{(2)}((k_1a)^j),$$

$$a_{n_1 p}^j = \frac{a_{n_{11p}}^j}{a_{n_{21}}^j}, \quad a_{n_1 q}^j = \frac{a_{n_{11q}}^j}{a_{n_{21}}^j},$$

where

$$a_{n_{11q}}^j = -(A_{n'n}^m)^{(2)} \psi_n((ka)^j)\xi_n^{'(1)}((ka)^j)\xi_n^{'(2)}((k_1a)^j)\xi_n^{(2)}((k_1a)^j) +$$

$$+(B_{n'n}^m)^2 \psi_n((ka)^j)\xi_n^{'(1)}((k_1a)^j)\xi_n^{'(2)}((ka)^j)\xi_n^{(2)}((k_1a)^j) +$$

$$+(B_{n'n}^m)^2 \psi_n'((ka)^j)(\xi_n^{'(2)}((k_1a)^j))\xi_n^{(1)}((ka)^j)\xi_n^{(1)}((ka)^j)\xi_n^{(2)}((k_1a)^j)$$

$$-(A_{n'n}^m)^2 \psi_n'((ka)^j)\xi_n^{(1)}((ka)^j)\xi_n^{'(2)}((k_1a)^j)\xi_n^{(2)}((k_1a)^j) +$$

$$+\left[(A_{n'n}^m)^2 - (B_{n'n}^m)^2\right] \psi_n((ka)^j)(\xi_n^{'(2)}((k_1a)^j))^2 (\xi_n^{(1)}((ka)^j))^2 -$$

$$-\left[(B_{n'n}^m)^2 - (A_{n'n}^m)^2\right] \psi_n'((ka)^j)\xi_n^{'(1)}((ka)^j)(\xi_n^{(2)}((k_1a)^j))^2 +$$

$$+(A_{n'n}^m)^2 \psi_n((ka)^j) F_{12}\xi_n^{'(1)}((ka)^j)\xi_n^{'(2)}((k_1a)^j) +$$

$$+(A_{n'n}^m)^2 \psi_n((ka)^j) F_{12} F_{21}((\xi_n^{'(1)}((k_1a)^j))^2 \xi_n^{(1)}((ka)^j) +$$

$$-(A_{n'n}^m)^2 \psi_n((ka)^j) F_{21}\xi_n^{(2)}((k_1a)^j)\xi_n^{'(1)}((k_1a)^j)\xi_n^{'(1)}((k1a)^j) +$$

$$-(A_{n'n}^m)^2 \psi_n((ka)^j) F_{12}\xi_n^{'(1)}((ka)^j)\xi_n^{(1)}((k_1a)^j)\xi_n^{'(2)}((k_1a)^j) +$$

$$+(B_{n'n}^m)^2 \psi_n((ka)^j) F_{21}\xi_n^{'(1)}((ka)^j)\xi_n^{(1)}((k_1a)^j)\xi_n^{(2)}((k_1a)^j) +$$

$$+(B_{n'n}^m)^2 \psi_n'((ka)^j) F_{12}\xi_n^{'(1)}((ka)^j)\xi_n^{'(1)}((k_1a)^j)\xi_n^{(2)}((k_1a)^j) -$$

$$-(B_{n'n}^m)^2 \psi_n((ka)^j) F_{21}\xi_n^{(1)}((ka)^j)\xi_n^{'(2)}((k_1a)^j)\xi_n^{'(1)}((k_1a)^j) -$$

$$-(B_{n'n}^m)^2 \psi_n((ka)^j) F_{12}\xi_n^{(1)}((k_1a)^j)\xi_n^{'(1)}((k_1a)^j)\xi_n^{'(2)}((k_1a)^j) +$$

$$+(B_{n'n}^m)^2 \psi_n((ka)^j) F_{12} F_{21}(\xi_n^{(1)}((ka)^j))(\xi_n^{'(1)}((k_1a)^j))^2 +$$

$$+(B_{n'n}^m)^2 \psi_n'((ka)^j) F_{21}\xi_n^{(2)}((k_1a)^j)\xi_n^{(2)}((k_1a)^j)\xi_n^{'(1)}((k_1a)^j) +$$

$$+(B_{n'n}^m)^2 \psi_n'((ka)^j) F_{12}\xi_n^{(1)}((ka)^j)\xi_n^{(1)}((k_1a)^j)\xi_n^{'(2)}((k_1a)^j) -$$

$$-(B_{n'n}^{m})^{2}\psi_{n}'((ka)^{j})F_{21}\xi_{n}^{(2)}((k_{1}a)^{j})\xi_{n}'^{(1)}((ka)^{j})\xi_{n}^{(1)}((k_{1}a)^{j})-$$

$$-(B_{n'n}^{m})^{2}\psi_{n}'((ka)^{j})F_{12}\xi_{n}^{(2)}((k_{1}a)^{j})\xi_{n}'^{(1)}((ka)^{j})\xi_{n}^{(1)}((k_{1}a)^{j})-$$

$$-(B_{n'n}^{m})^{2}\psi_{n}'((ka)^{j})F_{12}F_{21}\xi_{n}'^{(1)}((ka)^{j})(\xi_{n}^{(1)}((k_{1}a)^{j}))^{2}-$$

$$-(A_{n'n}^{m})^{2}\psi_{n}'((ka)^{j})F_{21}(\xi_{n}^{(1)}((ka)^{j}))(\xi_{n}'^{(2)}((ka)^{j}))\xi_{n}^{(1)}((k_{1}a)^{j})-$$

$$-(A_{n'n}^{m})^{2}\psi_{n}'((ka)^{j})F_{12}\xi_{n}^{(2)}((k_{1}a)^{j})\xi_{n}^{(1)}((ka)^{j})\xi_{n}^{(1)}((k_{1}a)^{j})+$$

$$+(A_{n'n}^{m})^{2}\psi_{n}'((ka)^{j})\xi_{n}^{(1)}((ka)^{j})\xi_{n}^{(1)}((k_{1}a)^{j})\xi_{n}^{(2)}((k_{1}a)^{j})F_{21}+$$

$$+(A_{n'n}^{m})^{2}\psi_{n}'((ka)^{j})(\xi_{n}^{(1)}((ka)^{j}))\xi_{n}^{(1)}((k_{1}a)^{j})\xi_{n}^{(2)}((k_{1}a)^{j})F_{12}+$$

$$+(A_{n'n}^{m})^{2}\psi_{n}'((ka)^{j})(\xi_{n}^{(1)}((k_{1}a)^{j}))^{2}\xi_{n}'^{(1)}((ka)^{j})F_{21}F_{12}+$$

$$+(A_{n'n}^{m})^{2}\psi_{n}((ka)^{j})(\xi_{n}'^{(2)}((k_{1}a)^{j}))\xi_{n}^{(1)}((ka)^{j})(\xi_{n}'^{(1)}((k_{1}a)^{j}))-$$

$$-(A_{n'n}^{m})^{2}\psi_{n}((ka)^{j})F_{12}F_{21}(\xi_{n}'^{(1)}((ka)^{j}))\xi_{n}'^{(1)}((k_{1}a)^{j})(\xi_{n}^{(1)}((k_{1}a)^{j}))+$$

$$(B_{n'n}^{m})^{2}\psi_{n}((ka)^{j})F_{12}F_{21}(\xi_{n}'^{(1)}((ka)^{j}))\xi_{n}'^{(1)}((k_{1}a)^{j})(\xi_{n}^{(1)}((k_{1}a)^{j}))-$$

$$a_{n_{11p}}^{j}=-(A_{n'n}^{m})(B_{n'n}^{m})F_{21}\psi_{n}'((ka)^{j})(\xi_{n}^{(2)}((k_{1}a)^{j}))^{2}\xi_{n}'^{(1)}((ka)^{j})\xi_{n}^{(1)}((ka)^{j})-$$

$$+(B_{n'n}^{m})^{2}F_{12}\psi_{n}'((ka)^{j})(\xi_{n}^{(1)}((ka)^{j}))\xi_{n}'^{(1)}((k_{1}a)^{j})\xi_{n}^{(1)}((k_{1}a)^{j})+$$

$$-(A_{n'n}^{m})^{2}F_{21}\psi_{n}'((ka)^{j})(\xi_{n}'^{(1)}((k_{1}a)^{j}))\xi_{n}^{(1)}((ka)^{j})\xi_{n}^{(1)}((k_{1}a)^{j})-$$

$$-(A_{n'n}^{m})(B_{n'n}^{m})F_{12}\psi_{n}((ka)^{j})(\xi_{n}'^{(1)}((k_{1}a)^{j}))\xi_{n}'^{(1)}((ka)^{j})\xi_{n}^{(2)}((k_{1}a)^{j})+$$

$$+(A_{n'n}^{m})(B_{n'n}^{m})F_{21}\psi_{n}'((ka)^{j})(\xi_{n}^{(1)}((ka)^{j}))\xi_{n}'^{(2)}((k_{1}a)^{j})\xi_{n}^{(1)}((k_{1}a)^{j})+$$

$$+(A_{n'n}^{m})(B_{n'n}^{m})F_{12}\psi_{n}'((ka)^{j})(\xi_{n}^{(1)}((ka)^{j}))\xi_{n}^{(2)}((k_{1}a)^{j})\xi_{n}'^{(1)}((k_{1}a)^{j}),$$

$$a_{n_{21}}^{j}=(-\xi_{n}^{(1)}((ka)^{j}))^{2}(A_{n'n}^{m})^{2}(\xi_{n}'^{(2)}((k_{1}a)^{j}))^{2}+F_{12}\xi_{n}'^{(1)}((k_{1}a)^{j})\xi_{n}^{(2)}((k_{1}a)^{j})+$$

$$+F_{21}\xi_{n}^{(1)}((k_{1}a)^{j})+((A_{n'n}^{m}))^{2}\xi_{n}^{(1)}((ka)^{j})\xi_{n}'^{(2)}((k_{1}a)^{j})+$$

$$+F_{12}\xi_{n}'^{(1)}((k_{1}a)^{j})\xi_{n}'^{(1)}((ka)^{j})\left[(\xi_{n}^{(2)}((k_{1}a)^{j})+\xi_{n}^{(1)}((k_{1}a)^{j})F_{21}\right]+$$

$$+\xi_{n}'^{(1)}((ka)^{j})(A_{n'n}^{m})^{2}\xi_{n}^{(1)}((ka)^{j})+F_{12}\xi_{n}'^{(1)}((k_{1}a)^{j})\xi_{n}^{(1)}((ka)^{j})\xi_{n}^{(2)}((k_{1}a)^{j})+$$

$$+F_{21}\xi_n^{'(1)}((ka)^j) - (A_{n'n}^m)^2\xi_n^{'(1)}((ka)^j)\xi_n^{(2)}((ka)^j)+$$

$$+F_{12}\xi_n^{'(1)}((ka)^j)\xi_n^{(1)}((k_1a)^j)\left[\xi_n^{(2)}((k_1a)^j) + \xi_n^{(1)}((k_1a)^j)F_{21}\right]-$$

$$-\xi_n^{(1)}((ka)^j)(B_{n'n}^m)^2\xi_n^{'(1)}((k_1a)^j) + F_{12}\xi_n^{'(1)}((ka)^j)\xi_n^{'(1)}((k_1a)^j)\xi_n^{(2)}((k_1a)^j)+$$

$$+F_{21}\xi_n^{(1)}((k_1a)^j) + (B_{n'n}^m)^2\xi_n^{'(1)}((k_1a)^j)\xi_n^{(1)}((ka)^j)^2 + +F_{12}\xi_n^{'(1)}((k_1a)^j)\xi_n^{'(2)}((ka)^j)^2+$$

$$+F_{21}\xi_n^{'(1)}((k_1a)^j) + (B_{n'n}^m)^2\xi_n^{'(1)}((ka)^j)+$$

$$+F_{12}\xi_n^{(1)}((k_1a)^j)\xi_n^{'(1)}((ka)^j)\xi_n^{(2)}((k_1a)^j) + F_{21}\xi_n^{(1)}((k_1a)^j)-$$

$$-\xi_n^{'(1)}((ka)^j)(B_{n'n}^m)^2 + F_{12}\xi_n^{(1)}((k_1a)^j)\xi_n^{(1)}((ka)^j)\xi_n^{'(2)}((k_1a)^j) + +F_{21}\xi_n^{'(1)}((k_1a)^j),$$

where

$$F_{12} = \frac{m_1^j\xi_n^{'(2)}((k_1b)^j)\psi_n((k_2b)^j) - m_2^j\xi_n^{(2)}((k_1b)^j)\psi_n'((k_2b)^j)}{m_2^j\xi_n^{(1)}((k_1b)^j)\psi_n'((k_2b)^j) - m_1^j\xi_n^{'(1)}((k_1b)^j)\psi_n((k_2b)^j)},$$

$$F_{21} = \frac{m_2^j\xi_n^{'(2)}((k_1b)^j)\psi_n((k_2b)^j) - m_1^j\xi_n^{(2)}((k_1b)^j)\psi_n'((k_2b)^j)}{m_1^j\xi_n^{(1)}((k_1b)^j)\psi_n'((k_2b)^j) - m_2^j\xi_n^{'(1)}((k_1b)^j)\psi_n((k_2b)^j)},$$

$\psi_n(\rho) = \rho j_n(\rho)$, $\xi_n^{(1)}(\rho) = \rho h_n^{(1)}(\rho)$, $\xi_n^{(2)}(\rho) = \rho h_n^{(2)}(\rho)$- are RiccatiBessel functions; prime denotes differentiation, $k = kn_o$, $k_1 = km_1^j$, $k_2 = km_2^j$, m_1^j, m_2^j are and nucleus, respectively, n_o is refractive index of the medium, a^j is the , b^j is the radius cytoplasm of the jth particle k is the wavenumber.

For a body of revolution, the vector spherical harmonics are represented as [9]:

$$\mathbf{M}_{nm,2}^{(q)} = \sum_{n'=0}^{\infty} D_{m'}^{(nm)}\mathbf{M}_{nm,1}^{(q)}, \quad \mathbf{N}_{nm,2}^{(q)} = \sum_{n'=0}^{\infty} D_{m'}^{(nm)}\mathbf{N}_{nm,1}^{(q)},$$

$$D_{m'}^{(n,m)} = e^{[i(m'\alpha+m\gamma)]}\left[\frac{(n+m')!(n-m')!}{(n+m)!(n-m)!}\right]^{\frac{1}{2}} \times$$

$$\times \sum_\sigma \binom{n+m}{n-m'-\sigma}\binom{n-m}{\sigma}(-1)^{n+m-\sigma} \times$$

$$\times \left[\cos\left[\frac{\beta}{2}\right]\right]^{2\sigma+m'+m} \left[\sin\left[\frac{\beta}{2}\right]\right]^{2n-2\sigma-m-m} ,$$

where α, β γ are Euler angles.

3.4 Scattering by a Group of Spherical Objects

The electromagnetic field that is incident on the surface of the jth particle consists of two parts: original incident field and the field that is scattered by a group of particles in a medium with refractive index N. Then, the following expressions are valid [3]:

$$\mathbf{E_i}(j) = \mathbf{E_0}(j) + \sum_{l \neq j} \mathbf{E_s}(l, j), \tag{3.56}$$

$$\mathbf{H_i}(j) = \mathbf{H_0}(j) + \sum_{l \neq j} \mathbf{H_s}(l, j), \tag{3.57}$$

where $\mathbf{E_s}(l, j)$ and $\mathbf{H_s}(l, j)$ are the sums of fields scattered at the jth particle. Indices l and j imply the transfer from the l to j coordinate system. The incident wave is defined in the following way:

$$\mathbf{E_{0(j)}} = -\sum_{n=1}^{\infty} \sum_{m=-n}^{n} i\, E_{mn}[p_{mn}^{j,j}\mathbf{N}_{mn}^1 + q_{mn}^{j,j}\mathbf{M}_{mn}^1], \tag{3.58}$$

$$\mathbf{H_{0(j)}} = -\frac{k}{\omega\mu}\sum_{n=1}^{\infty} \sum_{m=-n}^{n} i\, E_{mn}[q_{mn}^{j,j}\mathbf{N}_{mn}^1 + p_{mn}^{j,j}\mathbf{M}_{mn}^1] \tag{3.59}$$

The incident waves are considered in respect to the center of the jth particle, i.e., in the jth coordinate system.

The orientation of the wave vector $\mathbf{k}$ at an angle α to the axis z is defined as

$$\mathbf{k} = k(\mathbf{e_x} \sin\alpha \cos\beta + \mathbf{e_y} \sin\alpha \sin\beta + \mathbf{e_z} \cos\alpha),$$

where β is the angle between the axis x and the vector $\mathbf{k}$ component in the plane xy, and α is the angle of wave incidence in respect to the axis z. Usually, two polarizations of the incident wave are considered; i.e., p- and s-polarizations. For definiteness, we consider the p polarization. In this case, coefficients $p_{mn}^{j,j}$, $q_{mn}^{j,j}$ used in the expression for the incident field have the form

$$p_{mn}^{j,j} = \exp[i\mathbf{k} \cdot \mathbf{r}_{j_0,j}]p_{mn}^0, \quad q_{mn}^{j,j} = \exp[i\mathbf{k} \cdot \mathbf{r}_{j_0,j}]q_{mn}^0$$

where

$$q_{mn}^0 = \frac{1}{n(n+1)}\left[\frac{\partial}{\partial\alpha}P_n^m(\cos\alpha)\cos\beta - i\frac{m}{\sin\alpha}P_n^m(\cos\alpha)\sin\beta\right],$$

$$p_{mn}^0 = \frac{1}{n(n+1)}\left[\frac{\partial}{\partial\alpha}P_n^m(\cos\alpha)\cos\beta - i\frac{m}{\sin\alpha}P_n^m(\cos\alpha)\sin\beta\right].$$

In order to describe the scattering at an jth particle, we use the summation theorems for the vector-spherical functions [7]:

$$\mathbf{M}_{mn} = \sum_{\nu=0}^{\infty}\sum_{\mu=-\nu}^{\nu}[AO_{\mu\nu}^{mn}\mathbf{M}_{\mu\nu}' + BO_{\mu\nu}^{mn}\mathbf{N}_{\mu\nu}'],$$

$$\mathbf{N}_{mn} = \sum_{\nu=0}^{\infty}\sum_{\mu=-\nu}^{\nu}[BO_{\mu\nu}^{mn}\mathbf{M}_{\mu\nu}' + AO_{\mu\nu}^{mn}\mathbf{N}_{\mu\nu}'].$$

$\mathbf{M}_{mn}, \mathbf{N}_{mn}$ are the basis vector-spherical wave functions defined with the center at the point O, $\mathbf{M}_{\mu\nu}', \mathbf{N}_{\mu\nu}'$ are similar functions with the center at the point O', $\mathbf{M}_{\mu\nu}'$ and $\mathbf{N}_{\mu\nu}'$ have the form identical to $\mathbf{M}_{mn}, \mathbf{N}_{mn}$. Here,

$$AO_{\mu\nu}^{mn}(l,j) = (-1)^\mu i^{\nu-n}\frac{2\nu+1}{2\nu(\nu+1)}\sum_{p=|n-\nu|}^{n+\nu}(-i)^p[n(n+1)+$$

$$+\nu(\nu+1) - p(p+1)]\alpha(m,n,-\mu,\nu,p)h_p^1(kr_{(l,j)})\times$$

$$\times P_p^{m-\mu}(\cos\theta_{(l,j)})e^{i(m-\mu)\phi_{l,j}}, \tag{3.60}$$

$$BO_{\mu\nu}^{mn}(l,j) = (-1)^\mu i^{\nu-n}\frac{2\nu+1}{2\nu(\nu+1)}\sum_{p=|n-\nu|}^{n+\nu}(-i)^p b[m,n,-\mu,\nu,p,p-1]\times$$

$$\times h_p^1(kr_{(l,j)})P_p^{m-\mu}(\cos\theta_{(l,j)})e^{i(m-\mu)\phi_{l,j}}, \tag{3.61}$$

$$b(m,n,-\mu,\nu,p,p-1) = \frac{2p+1}{2p-1}[(\nu-\mu)(\nu+\mu+1)]\times$$

$$\times\frac{2p+1}{2p-1}[\alpha(m,n,-\mu-1,\nu,p-1) - (p-m+\mu)(p-m+\mu-1)]\times$$

$$\times\frac{2p+1}{2p-1}[\alpha(m,n,-\mu+1,\nu,p-1) + 2\mu(p-m+\mu)]\times$$

$$\times \frac{2p+1}{2p-1} \alpha(m, n, -\mu, \nu, p-1),$$

where

$$\alpha(m, n, \mu, \nu, p) = \frac{2p+1}{2} \frac{(p-m-\mu)!}{(p+m+\mu)!} \int_{-1}^{1} P_n^m(x) P_\nu^\mu(x) P_p^{m+\mu}(x) dx.$$

In these expressions $r_{l,j}, \theta_{l,j}, \phi_{l,j}$ are the spherical coordinates at the center of the lth particle in the jth coordinate system. From the summation theorems it follows that

$$\mathbf{M}_{mn}^3(l) = \sum_{\nu=0}^{\infty} \sum_{\mu=-\nu}^{\nu} [A O_{\mu\nu}^{mn}(l, j) \mathbf{M}_{\mu\nu}^1(j) + B O_{\mu\nu}^{mn}(l, j) \mathbf{N}_{\mu\nu}^1(j)], \qquad (3.62)$$

$$\mathbf{N}_{mn}^3(l) = \sum_{\nu=0}^{\infty} \sum_{\mu=-\nu}^{\nu} [B O_{\mu\nu}^{mn}(l, j) \mathbf{M}_{\mu\nu}^1(j) + A O_{\mu\nu}^{mn}(l, j) \mathbf{N}_{\mu\nu}^1(j)]. \qquad (3.63)$$

Let us derive expressions for the scattered field from (3.58) taking into account (3.62) and (3.63); we obtain

$$\mathbf{E}_s(l, j) = -\sum_{n=1}^{\infty} \sum_{m=-n}^{n} i \, E_{mn} [p_{mn}^{l,j} \mathbf{N}_{mn}^1 + q_{mn}^{l,j} \mathbf{M}_{mn}^1], \qquad (3.64)$$

$$\mathbf{H}_s(l, j) = -\frac{k}{\mu\omega} \sum_{n=1}^{\infty} \sum_{m=-n}^{n} E_{mn} [q_{mn}^{l,j} \mathbf{N}_{mn}^1 + p_{mn}^{l,j} \mathbf{M}_{mn}^1], \qquad (3.65)$$

where

$$p_{mn}^{l,j} = -\sum_{\nu=1}^{\infty} \sum_{\mu=-\nu}^{\nu} [a_{\mu\nu}^l A_{mn}^{\mu\nu}(l, j) + b_{\mu\nu}^l B_{mn}^{\mu\nu}(l, j)],$$

$$q_{mn}^{l,j} = -\sum_{\nu=1}^{\infty} \sum_{\mu=-\nu}^{\nu} [a_{\mu\nu}^l B_{mn}^{\mu\nu}(l, j) + b_{\mu\nu}^l A_{mn}^{\mu\nu}(l, j)],$$

$$A_{mn}^{\mu\nu} = \frac{E_{\mu\nu}}{E_{mn}} A O_{mn}^{\mu\nu} = i^{\nu-n} \frac{(2\mu+1)(n+m)!(\nu-\mu)!}{(2n+1)(n-m)!(\nu+\mu)!} A O_{mn}^{\mu\nu},$$

$$B_{mn}^{\mu\nu} = \frac{E_{\mu\nu}}{E_{mn}} B O_{mn}^{\mu\nu} = i^{\nu-n} \frac{(2\mu+1)(n+m)!(\nu-\mu)!}{(2n+1)(n-m)!(\nu+\mu)!} B O_{mn}^{\mu\nu}.$$

Substituting (3.8), (3.59), (3.64) into (3.58), we obtain

$$p_{mn}^j = p_{mn}^{j,j} - \sum_{l \neq j}^{L} \sum_{v=1}^{\infty} \sum_{\mu=-v}^{v} [a_{\mu v}^l A_{mn}^{\mu v}(l, j) + b_{\mu v}^l B_{mn}^{\mu v}(l, j)], \tag{3.66}$$

$$q_{mn}^j = q_{mn}^{j,j} - \sum_{l \neq j}^{L} \sum_{v=1}^{\infty} \sum_{\mu=-v}^{v} [a_{\mu v}^l B_{mn}^{\mu v}(l, j) + b_{\mu v}^l A_{mn}^{\mu v}(l, j)]. \tag{3.67}$$

The system of linear algebraic equations that makes it possible to find coefficients and with allowance for multiple scattering for j particles with nonconcentric inclusions:

$$a_{mn}^j = a_{n_{1p}}^j [p_{mn}^{j,j} - \sum_{l \neq j}^{L} \sum_{v=1}^{\infty} \sum_{\mu=-v}^{v} [a_{\mu v}^l A_{mn}^{\mu v}(l, j) + b_{\mu v}^l B_{mn}^{\mu v}(l, j)]] +$$

$$+ a_{n_{1q}}^j [q_{mn}^{j,j} - \sum_{l \neq j}^{L} \sum_{v=1}^{\infty} \sum_{\mu=-v}^{v} [a_{\mu v}^l B_{mn}^{\mu v}(l, j) + b_{\mu v}^l A_{mn}^{\mu v}(l, j)]],$$

$$b_{mn}^j = b_{n_{1q}}^j [q_{mn}^{j,j} - \sum_{l \neq j}^{L} \sum_{v=1}^{\infty} \sum_{\mu=-v}^{v} [a_{\mu v}^l A_{mn}^{\mu v}(l, j) + b_{\mu v}^l B_{mn}^{\mu v}(l, j)]] + \tag{3.68}$$

$$+ b_{n_{1p}}^j [p_{mn}^{j,j} - \sum_{l \neq j}^{L} \sum_{v=1}^{\infty} \sum_{\mu=-v}^{v} [a_{\mu v}^l B_{mn}^{\mu v}(l, j) + b_{\mu v}^l A_{mn}^{\mu v}(l, j)]],$$

$$n = 1, 2, 3, 4, 5, \ldots \ , m = 0, 1, 2, 3, 4, 5, \ldots, n$$

A matrix representation of this system of equations is written as

$$\begin{pmatrix} a^j \\ b^j \end{pmatrix} = T_1^j \left[\begin{pmatrix} p^{j,j} \\ q^{j,j} \end{pmatrix} + \sum_{l \neq j} \begin{pmatrix} A(l, j) & B(l, j) \\ B(l, j) & A(l, j) \end{pmatrix} \begin{pmatrix} a^j \\ b^j \end{pmatrix} \right] + \tag{3.69}$$

$$\left(T_2^j \right) \left[\begin{pmatrix} p^{j,j} \\ q^{j,j} \end{pmatrix} + \sum_{l \neq j} \begin{pmatrix} A(l, j) & B(l, j) \\ B(l, j) & A(l, j) \end{pmatrix} \begin{pmatrix} a^j \\ b^j \end{pmatrix} \right],$$

or

$$\begin{pmatrix} a^j \\ b^j \end{pmatrix} = T_{12}^j \left[\begin{pmatrix} p^{j,j} \\ q^{j,j} \end{pmatrix} + \sum_{l \neq j} \begin{pmatrix} A(l, j) & B(l, j) \\ B(l, j) & A(l, j) \end{pmatrix} \begin{pmatrix} a^j \\ b^j \end{pmatrix} \right], \tag{3.70}$$

$$T_{12}^j = T_1^j + T_2^j, \quad T_1^j = \begin{pmatrix} a_{n_{1p}}^j & 0 \\ 0 & b_{n_{1q}}^j \end{pmatrix}, \quad T_2^j = \begin{pmatrix} 0 & a_{n_{1q}}^j \\ b_{n_{1p}}^j & 0 \end{pmatrix},$$

System (3.70) must be solved with the aid of the method of reduction in which a finite number of equations and a finite number of unknown quantities are considered and a stable algorithm of biconjugate gradients [10, 11] is employed.

When coefficients a_{mn}^j and b_{mn}^j are found using system (3.70), the expressions for the scattered field in the principal coordinate system can be written as

$$\mathbf{E}_s = \sum_{n=1}^{\infty} \sum_{m=-n}^{n} i \, E_{mn} [a_{mn}^j \mathbf{N}_{mn}^3 + b_{mn}^j \mathbf{M}_{mn}^3], \tag{3.71}$$

$$\mathbf{H}_s = \frac{k}{\omega\mu} \sum_{n=1}^{\infty} \sum_{m=-n}^{n} i \, E_{mn} [b_{mn}^j \mathbf{N}_{mn}^3 + a_{mn}^j \mathbf{M}_{mn}^3], \tag{3.72}$$

where

$$a_{mn}^j = \sum_{l=1}^{L} \sum_{v=1}^{\infty} \sum_{\mu=-v}^{v} [a_{\mu v}^l A_{mn}^{\mu v}(l, j_0) + b_{\mu v}^l B_{mn}^{\mu v}(l, j_0)],$$

$$b_{mn}^j = \sum_{l=1}^{L} \sum_{v=1}^{\infty} \sum_{\mu=-v}^{v} [a_{\mu v}^l B_{mn}^{\mu v}(l, j_0) + b_{\mu v}^l A_{mn}^{\mu v}(l, j_0)].$$

The system for the coefficients a_{mn}^j, b_{mn}^j can be simplified if we consider a part of the field that is forward-or backward-scattered by the particles at small angles relative to the Z axis.

The expressions for the scattered field in the far-field zone are represented as

$$E_{s\theta} \sim E_0 \frac{e^{ikr}}{-ikr} \sum_{n=1}^{\infty} \sum_{m=-n}^{n} (2n+1) \frac{(n-m)!}{(n+m)!} [a_{mn}^j \tau_{mn} + b_{mn}^j \pi_{mn}] e^{im\phi} \tag{3.73}$$

$$E_{s\phi} \sim E_0 \frac{e^{ikr}}{-ikr} \sum_{n=1}^{\infty} \sum_{m=-n}^{n} (2n+1) \frac{(n-m)!}{(n+m)!} [a_{mn}^j \pi_{mn} + b_{mn}^j \tau_{mn}] e^{im\phi}, \tag{3.74}$$

where

$$\tau_{mn} = \frac{\partial}{\partial\theta} P_n^m(\cos\theta), \pi_{mn} = \frac{m}{\sin\theta} P_n^m(\cos\theta)$$

Symbol ($\sim$) indicates asymptotic interpretation of expressions (3.73) and (3.74), which follow from expression (3.71) at ($kr \gg 1$). We consider the scattering at relatively large distances from the jth particle. Therefore, the electric vectors of the scattered field are parallel to the electric field of the incident field, so that only the component differs from zero in the far-field zone. Note that we consider a part of the scattered field inside the cavity.

Expressions (3.73) and (3.74) can be simplified:

$$E_{s\theta} \sim E_0 \frac{e^{ikr}}{-ikr} \sum_{n=1}^{\infty} \sum_{m=-n}^{n} \frac{(2n+1)}{n(n+1)} [a_{mn}^j \tau_n + b_{mn}^j \pi_n]$$

$$E_{s\phi} \sim E_0 \frac{e^{ikr}}{-ikr} \sum_{n=1}^{\infty} \sum_{m=-n}^{n} \frac{(2n+1)}{n(n+1)} [a_{mn}^j \pi_n + b_{mn}^j \tau_n], \qquad (3.75)$$

where

$$\tau_n = \frac{\partial}{\partial \theta} P_n(\cos\theta), \pi_n = \frac{1}{\sin\theta} P_n(\cos\theta)$$

Similar expressions can be derived for magnetic field H.

3.5 Numerical Study of the Algebraic Equations

To solve problems of light scattering of dielectric bodies, simulating, for example, blood cells, it is often a problem solving ill-conditioned linear algebraic systems equations. Numerical analysis shows that the use of iterative methods is an effective step to solve linear algebraic systems with an ill-conditioned matrix. The most effective and stable method of iteration are projection methods, and particularly that of their class, which is associated with designing the Krylov subspace [10, 12].

Algorithm of Krylov subspace methods include two steps:
1. Construct a basis in the Krylov subspace.
2. The calculation of correcting the corrective

To calculate the correcting corrective amendments the following approaches were used

1. The Ritz–Galerkin method. The residual construction is orthogonal to Krylov's subspace. This approach is used in such methods, as a method of conjugate gradients and a method of a full orthogonalization.
2. The minimum residual approach. At each iteration, minimizing the norm of the residual. The approach used in the method of minimal residual for Krylov subspace and generalized minimal residual method.
3. The Petrov–Galerkin approach. The methods in this class are based on the construction of biorthogonal basis.

These methods have several advantages: they are stable, thanks to technology orthogonalization allowing efficient parallelization and work with different types of preconditioners, and these methods can be used for systems with nonsymmetric matrices. Thus, by solving a system of linear equations (3.68) a stable algorithm biconjugate gradient is used. This method is based on a quadratic conjugate gradient method, but does not allow the accumulation of rounding errors and unstable behavior of the residual [12]. The dependence of the relative residual norm on the number of iterations for the predetermined method of bioconjugate gradients (Fig. 3.2a).

The calculations showed that the use of iterative methods without additional modification is not reasonable, because in most cases the iterative methods have shown unsatisfactory convergence. Therefore, iterative methods have been used with preconditioning, as it improves the convergence to the desired solution [11].

We consider the following system linear algebraic equations

$$A\mathbf{x} = \mathbf{b} \tag{3.76}$$

where $\mathbf{b}$ is vector of free members, $\mathbf{x}$ is vector of unknowns and A is matrix $(N \times N)$ coefficients of the system.

Let M be is nonsingular matrix $(N \times N)$. Multiplying (3.76) by the matrix M^{-1}, we obtain the system

$$M^{-1}A\mathbf{x} = M^{-1}\mathbf{b}, \tag{3.77}$$

The system is same exact solution x_*, because M is nonsingular matrix.

The process transition from (3.76) to (3.77) for the purpose of improving the characteristics of the matrix to accelerate the convergence of the solution is called preconditioning, and the matrix M^{-1} is matrix preconditioner. Methods of preconditioning can be divided into two types: explicit and implicit. The preconditioning can

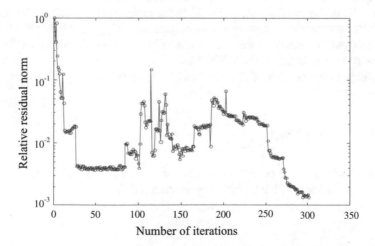

Fig. 3.3 The dependence of the relative residual norm on the number of iterations for the predetermined method of bioconjugate gradients for the following parameters: the number of particles in the layer being simulated was assumed to be ten, the relative refractive index is $1.035 + 0.00001i$, radius of the particle is $3.5 \, \mu m$

Fig. 3.4 Dependence of the relative residual norm on the number of iterations for the predetermined method of bioconjugate gradients for the following parameters: the number of particles in the layer being simulated was assumed to be ten, the relative refractive index is $1.035 + 0.00001i$, radius of the particle is 3.5 μm

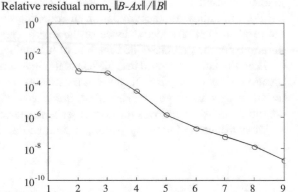

be introduced into the scheme of the method without the need for explicit calculation of the matrix product.

Thus, an explicit preconditioning requires finding the matrix M^{-1} and matrix multiplication preconditioning on vector in each iteration. If we consider the implicit method for solving linear algebraic equation, in this case it is necessary to solve linear algebraic equation with the matrix M in each iteration.

The majority of methods in both types of preconditioning is based on the representation of the product of two matrices L and U, i.e. $M = LU$ (LU is decomposition). We solve the linear algebraic equation with the preconditioner in the form of LU-decomposition. Figures 3.3 and 3.4 shows the relative residual norm of the iteration number for preconditioned methods of bioconjugate gradients and indicates sufficient convergence of the method. From the graphs follows the conclusion of a sufficiently stable convergence of the method (Fig. 3.5).

The array of the number of iterations has a step of 0.1.

3.6 Scattering Matrix

In what follows, we will need the elements of the scattering matrix, which relates the Stokes parameters of the incident and scattered fields

$$\mathbf{L}_s = S\mathbf{L}_i,$$

where $\mathbf{L}_i$ is the Stokes vector of the incident field, $\mathbf{L}_s$ is the Stokes vector of the scattered field, and S is the 4×4 scattering matrix. Elements of this matrix are expressed in terms of the elements of the 2×2 matrix that relates the orthogonal components of electric vectors of the scattered ($\mathbf{E}_{\parallel s}$, $\mathbf{E}_{\perp s}$) and incident ($\mathbf{E}_{\parallel i}$, $\mathbf{E}_{\perp i}$) electromagnetic waves,

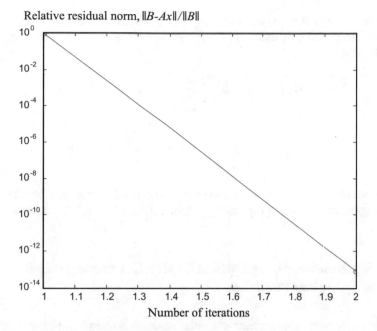

Fig. 3.5 Dependence of the relative residual norm on the number of iterations for the predetermined method of bioconjugate gradients for the following parameters: the number of particles in the layer being simulated was assumed to be ten, the nucleus diameter is 4×10^{-6} m; the diameter of the cytoplasm is 6×10^{-6} m and blood plasma are, respectively $1.31 + j0.0001$, $1.32 + j0.0001$, and 1.52, $d = 1.6 \times 10^{-7}$ m

$$\begin{pmatrix} \mathbf{E}_{\parallel s} \\ \mathbf{E}_{\perp s} \end{pmatrix} = \begin{pmatrix} \mathbf{E}_{s\theta} \\ -\mathbf{E}_{s\phi} \end{pmatrix} = \frac{e^{ikr-ikz}}{-ikr} \begin{pmatrix} S_2 & S_3 \\ S_4 & S_1 \end{pmatrix} \begin{pmatrix} E_{\parallel i} \\ E_{\perp i} \end{pmatrix}. \tag{3.78}$$

To write the field scattering forward (backward) in the small-angle vicinity of the wave propagation direction, it is sufficient to use the diagonal representation of the scattering matrix S

$$S = \begin{pmatrix} S_{11} & 0 & 0 & 0 \\ 0 & S_{22} & 0 & 0 \\ 0 & 0 & S_{33} & 0 \\ 0 & 0 & 0 & S_{44} \end{pmatrix},$$

where

$$S_{11} = \frac{1}{2}[|S_2|^2 + |S_1|^2] = S_{22},$$

$$S_{33} = \frac{1}{2}[S_1 S_2^* + S_2 S_1^*] = S_{44}.$$

Here the sign (*) denotes the complex conjugation, and the expressions for the scattering amplitudes S_1 and S_2 of the passed ($\theta = 0$) and reflected ($\theta = \pi$) waves have the form

$$S_2(0) = S_1(0) = \frac{1}{2} \sum_{n=1}^{\infty} \sum_{m=-n}^{n} (2n + 1)[a_{mn} + b_{mn}], \qquad (3.79)$$

$$S_2(\pi) = -S_1(\pi) = \frac{1}{2} \sum_{n=1}^{\infty} \sum_{m=-n}^{n} (2n + 1)(-1)^n [a_{mn} - b_{mn}]. \qquad (3.80)$$

Expressions (3.79) and (3.80) will be subsequently used to calculate the frequencies of the eigenmodes in the optical cavity with an ensemble of spherical particles.

3.7   Eigenmodes of the Optical Cavity Containing a Cell of Spherical Particles

Since the eigenmodes of annular and linear resonators change differently when an inhomogeneous medium is introduced into them, for definiteness we will consider the simplest, linear resonator. The resonator scheme is shown in Fig. 3.1.

We assume that the plane of the resonator optical contour is the symmetry plane. This assumption is necessary to justify both the subsequent separation of variables in the field equations and the small degree of depolarization of the field transmitted through the layer of spherical particles. The closed system of equations for the field E in a cross section orthogonal to the optical contour of a two-mirror resonator can be written, analogously to [13], in the form

$$E^{\pm} = (I + R_1 R_2)E^{\pm}, \qquad (3.81)$$

where I is the matrix operator describing eigenmodes of the cavity without the medium, and R_1, R_2 are the same for the cavity with the medium. After the separation of variables in (3.81), the expanded integral equation for a coordinate cofactor of the scalar component U of the eigenmode field at a resonator mirror has the form

$$U(\xi) = \sqrt{\frac{\mp i}{2\pi B}} \exp\left[\pm ikL\right] \int_{-\infty}^{\infty} \exp\left[\ln[R_1(x_1)R_2(x_1)]\right] \times \qquad (3.82)$$

$$\times \exp\left[(\pm i(Ax_1^2 + D\xi^2 - 2\xi x_1)/(2B)\right] U(x_1)dx_1,$$

where the signs ($-$) and ($+$) correspond to the field at the left- and right-hand mirrors, respectively,? R_1 and R_2 are the scalar functions of the form

$$R_1 = S(0) \exp\left[-\frac{i}{z_{11}} + \frac{1}{2}\frac{x_1^2}{z_{11}^2}\right], \; R_2 = S(\pi) \exp\left[\frac{i}{z_{22}} - \frac{1}{2}\frac{x_1^2}{z_{22}^2}\right], \quad (3.83)$$

$x_1 = \sqrt{k}x$, $z_{11} = z_1 k$, $z_{22} = z_2 k$, $S(0)$ and $S(\pi)$ are the scattering amplitudes for the transmitted and reflected waves, respectively, A, B, C and D are the wave matrix elements of the resonator without particles (effect of the particles is accounted for the coefficients R_1 and R_2), z_{11}, z_{22} are the distances from the particle layer along the resonator optical axis (the distances larger than these asymptotic formulas (3.75) for the scattered field are valid)

Expressions (3.83) are obtained by the Taylor expansion of the distance r used in (3.78) from the coordinate origin to the observation point on condition that

$$x^2/z^2 \ll 1$$

(small-angle assumption). The field distribution at the mirrors of the resonator with a particle layer in the domain Ω (see Fig. 3.1), is determined from the solution of integral equation (3.82) and has the form

$$U_n(x_1)^{\pm} = \frac{1}{2^n n! \varpi \pi} H_n\left(\frac{x_1}{\varpi}\right) \times$$

$$\times \exp\left[\mp i(n + 1/2)\tilde{g} + (\varepsilon - \delta) + \ln(\rho_1) \pm ikL + \ln[S(0)S(\pi)] \pm \frac{ix_1^2}{4}\right].$$

Therewith, the resonator eigenmodes are expressed by the formula

$$\omega_n = c\frac{2\pi q + (n + 1/2)\tilde{g} - i((\varepsilon - \delta) + \ln(\rho_1) + \ln[S(0)S(\pi)])}{LN}, \quad (3.84)$$

where c is the speed of light in vacuum, q is the number of longitudinal mode, n is the number of transverse mode $q \gg n$, L is the resonator length, N is the environment refractive index,

$$\varpi = \sqrt{\frac{\sin \tilde{g}}{B}}, \; \tilde{g} = \arccos\left[\frac{\tilde{A} + D}{2}\right], \; \varepsilon = \frac{i}{z_{22}}, \; \delta = -\frac{i}{z_{11}},$$

$$\tilde{A} = \left[A + \frac{i}{2z_{22}^2} - \frac{i}{2z_{11}^2}\right], \; \frac{1}{q} = \left[\frac{\tilde{A} + D}{2} + i\sqrt{1 - \frac{(\tilde{A} + D)^2}{4}} - \tilde{A}\right](2B)^{-1},$$

$\rho_1 = \rho k$ is the dimensionless thickness of the particle layer and H_n are the Hermitian polynomials.

Formula (3.84) implicitly defines the rather complicated relation between the frequencies of the resonator eigenmodes and the electrical parameters of the particles, such as the real and imaginary parts of their refractive indices, dimensions, etc. This formula cannot be further simplified without loss of information, and, hence, the relations between the mode frequencies and parameters of the medium and resonator should be analyzed numerically.

The model is implemented in a software package that allows automatic variations in the measured real and imaginary parts of the refractive indices and sizes of particles on a single setup. Such an approach makes it possible to reveal correlations of the electrophysical parameters and biological properties of blood formed elements. The optical parameters of blood cells must substantially supplement a detailed analysis of blood owing to more precise characterization of the cells. The model for the estimation of the refractive indices and sizes of blood formed elements and the intracavity measurements can be more informative and accurate in comparison with the existing methods that employ cavity-free models.

3.8 Numerical Calculations for the Resonator with a Simulated Medium and Conclusions

A very important area of application of laser radiation is biomedical optics. Using optical techniques, one can study mechanisms behind the interaction of cells with the environment and their response to changes in the physical properties of the medium and thereby gain information about an ensemble of living cells, including blood corpuscles (hemocytes), which play a key role in different physicochemical interactions. The advantage of laser radiation (laser beam) in studying biological particles is that it does not cause crude pathomorphological changes in the tissue. At the same time, laser diagnostics effectively utilizes such properties of laser radiation as coherency, monochromaticity, and directionality. Development of new methods of laser biomedical diagnostics requires a theoretical analysis of light propagation in biological tissues. The presence of an adequate theory will make it possible to better appreciate optical measurement data and raise the potential, reliability, and usefulness of optical technologies.

It is believed that such an approach will expand the information content of the intracavity method in analyzing the dependences of the optical characteristics on the radiation wavelength. In this work, we simulate the absorption versus the wavelength dependence for different parameters of the medium being simulated. The simulation demonstrates the feasibility of this method for studying biological structures of various configurations, such as an ensemble of spherulated particles with a nonconcentric inclusion (hemocyte suspensions). Erythrocytes have largely a spherical doubly concave shape. However, in general, the shape of an erythrocyte depends on intracellular factors and the environment. Sometimes, erythrocytes may be spherical, for example, when the cell is in a hypotonic solution.

A mathematical model was constructed and the respective software was devised that allowed us to perform a numerical experiment at different parameters of the problem. Some of the results are presented below.

Consider a layer of spherical particles in an optical resonator cavity. Its parameters are the following: the mirror distance is $L = 11$ cm; the radii of mirrors M_1 and M_2 are 100.0 and 46.3 cm, respectively; the radiation wavelength of a helium-neon laser is $\lambda = 0.63\,\mu$m; $z_1 = 6000\lambda$; and $z_2 = 6005\lambda$. Thickness ρ of the layer is set equal to the diameter of the particles, the distance between which in the plane orthogonal to the optical axis of the cavity is taken to be equal to 10λ, From formula (3.84), one can calculate the refractive index of the particles, $n^0(\lambda) + i\chi(\lambda)$ at given cavity frequency ω_n. In the given case, we put n = 0, so that ω_0 is the frequency of the fundamental mode.

Experiments were carried out with spherulated particles containing a nonconcentric inclusion. The number of spherulated particles was set equal to five. Let the parameters of the spherulated particles be the following: the nucleus diameter is 4 $\times 10^{-6}$m; the diameter of the cytoplasm is 6×10^{-6}m; and the refractive indices of the nucleus, cytoplasm, and blood plasma are, respectively 1.31 + j0.0001, 1.32 + j0.0001, and 1.52.

Some of the hemocytes have a nucleus. The nucleus is not always placed at the center. It may be spherical, ovoid, etc. It seems topical to study the spectral response of various biological samples to a change in the nucleus position under laser irradiation.

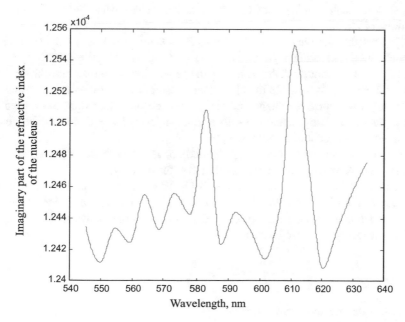

Fig. 3.6 Imaginary part of the refractive index of the leukocyte nucleus versus the wavelength for $d = 1.6 \cdot 10^{-7}$

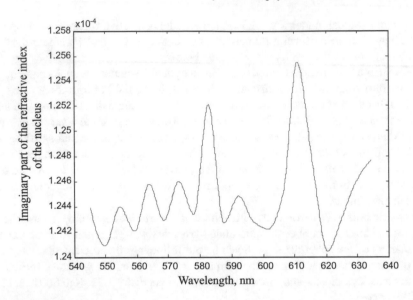

Fig. 3.7 Imaginary part of the refractive index of the leukocyte nucleus versus the wavelength for $d = 1.8 \cdot 10^{-7}$

Figures 3.6 and 3.7 plot the imaginary (χ) part of the refractive index of hemocytes with a nonconcentric inclusion of a different radius versus the wavelength for different positions of the nucleus.

It is seen that the range of χ is close to the experimental range of the complex refractive index of the hemocyte nucleus obtained without using the intracavity model [14, 15]. Also, the model is fairly sensitive to the position of the nucleus relative to the center. This makes it possible to gain a deeper insight into physiological processes in the organism, since the shape and size of the nucleus may vary, often together with metabolism changes, and the shift of the nucleus may be due to the hemocyte damage or impairment of a hemocyte [16].

The same dependences may be simulated for lasers with other parameters and used to process experimental data, specifically, for hemocytes.

Figures 3.8, 3.9, 3.10, 3.11, 3.12 and 3.13 illustrate the cross section of multiple scattering by a set of spherulated particles with a nonconcentric inclusion of a different radius in the far-field region for different positions of the nucleus.

The scattering cross section is given by

$$C_{sca} = \frac{W_{scat}}{I_i},$$

where I_i is the incident light intensity

$$W_{scat} = \int_A S_{scat} \cdot \mathbf{e_r} dA, \quad S_{scat} = \frac{c}{8\pi} \mathrm{Re}[\mathbf{E}_{scat}^j \times \mathbf{H}_{scat}^{j*}],$$

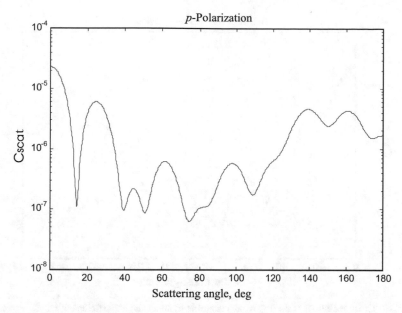

Fig. 3.8 Cross section of scattering by an ensemble of spherulated particles with a nonconcentric inclusion at $d = 1.7 \cdot 10^{-7}$ m for p-polarized

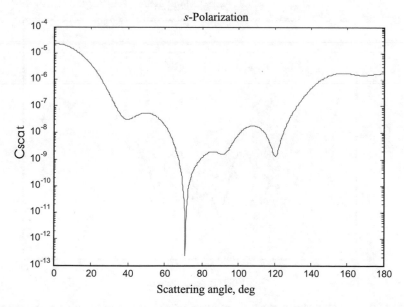

Fig. 3.9 Cross section of scattering by an ensemble of spherulated particles with a nonconcentric inclusion at $d = 1.7 \cdot 10^{-7}$ m for s-polarized

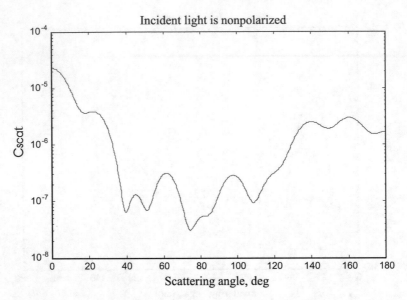

Fig. 3.10 Cross section of scattering by an ensemble of spherulated particles with a nonconcentric inclusion at $d = 1.7 \cdot 10^{-7}$ m for nonpolarized incident light

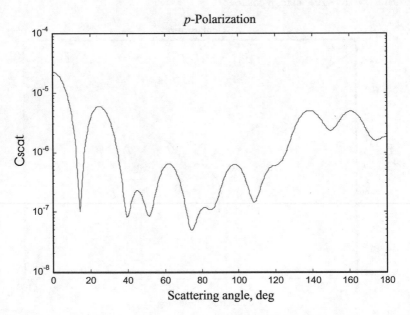

Fig. 3.11 Cross section of scattering by an ensemble of spherulated particles with a nonconcentric inclusion of a different radius at $d = 1.687 \cdot 10^{-7}$ m for p-polarized

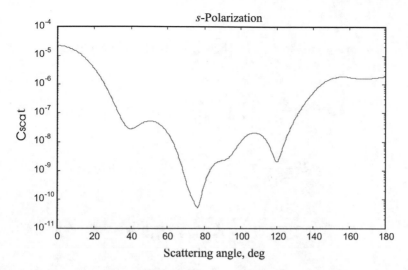

Fig. 3.12 Cross section of scattering by an ensemble of spherulated particles with a nonconcentric inclusion of a different radius at $d = 1.687 \cdot 10^{-7}$ m for s-polarized

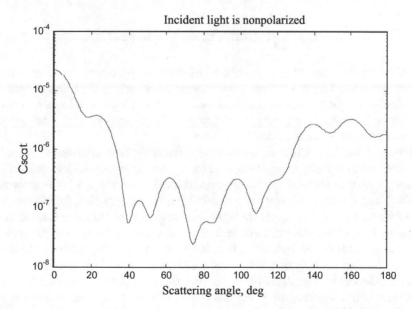

Fig. 3.13 Cross section of scattering by an ensemble of spherulated particles with a nonconcentric inclusion of a different radius at $d = 1.687 \cdot 10^{-7}$ m for nonpolarized incident light

Then,

$$W_{scat} = \frac{1}{2}\text{Re} \int_0^{2\pi} \int_0^{\pi} \left[\mathrm{E}_{s\theta}\mathrm{H}_{s\phi}^* - \mathrm{E}_{s\phi}\mathrm{H}_{s\theta}^* \right] r^2 \sin\theta \mathrm{d}\theta \mathrm{d}\phi,$$

where

$$E_{s\theta} \sim E_0 \frac{e^{ikr}}{-ikr} \sum_{n=1}^{\infty} \sum_{m=-n}^{n} \frac{(2n+1)}{n(n+1)} [a_{mn}\tau_n + b_{mn}\pi_n],$$

$$E_{s\phi} \sim E_0 \frac{e^{ikr}}{-ikr} \sum_{n=1}^{\infty} \sum_{m=-n}^{n} \frac{(2n+1)}{n(n+1)} [a_{mn}\pi_n + b_{mn}\tau_n],$$

$$H_{s\theta} \sim E_0 \frac{e^{ikr}}{-ikr} \frac{k}{\omega\mu} \sum_{n=1}^{\infty} \sum_{m=-n}^{n} \frac{(2n+1)}{n(n+1)} [b_{mn}\tau_n + a_{mn}\pi_n],$$

$$H_{s\phi} \sim E_0 \frac{e^{ikr}}{-ikr} \frac{k}{\omega\mu} \sum_{n=1}^{\infty} \sum_{m=-n}^{n} \frac{(2n+1)}{n(n+1)} [b_{mn}\pi_n + a_{mn}\tau_n],$$

$$\tau_n = \frac{\partial}{\partial\theta} P_n(\cos\theta), \pi_n = \frac{1}{\sin\theta} P_n(\cos\theta).$$

Of undeniable interest is to study the polarization characteristics of an ensemble of spherulated particles with a nonconcentric inclusion. It is known that the polarization characteristics of scattered radiation contain more information about the microstructure and also the geometrical and optical properties of scatterers and serve as a sensitive tool in optical diagnostics.

Figures 3.14, 3.15, 3.16 and 3.17 plot the polarization characteristics of the radiation scattered by hemocyte suspensions against the scattering angle. Here, normalized Stokes parameters U and V were determined through S_{33} and S_{34}, which are the normalized components of the scattering matrix for spherical particles. As follows from the plots, the Stokes parameters are highly sensitive not only to the refractive index of the particles with a nonconcentric inclusion but also to the position of the nucleus. It is also seen that the interval $60° < \theta < 90°$ (θ is the scattering angle) is of interest for experimental investigation, since here oscillations are the least pronounced.

Thus, the model suggested for estimating the absorption factor of hemocytes combined with an intracavity experiment may, in our opinion, be more adequate than the available methods based on cavity-free models. An undeniable advantage of our approach is that data for the imaginary part of the refractive index, hemocyte size, and other parameters can be obtained simultaneously on the same setup. The model allows one to determine the spectral distributions of the optical parameters of a biological medium and trace the variation of these parameters under the action of various factors causing changes in the functional and morphological state of a biological tissue. In addition, using this model, one can simultaneously gain data for the variation of the optical parameters and characteristic sizes of biological tissues

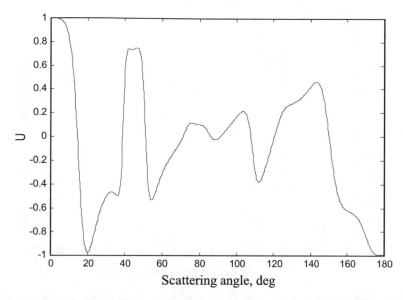

Fig. 3.14 Stokes parameters versus the scattering angle for the hemocyte suspension models at $d = 1.7 \cdot 10^{-7}$

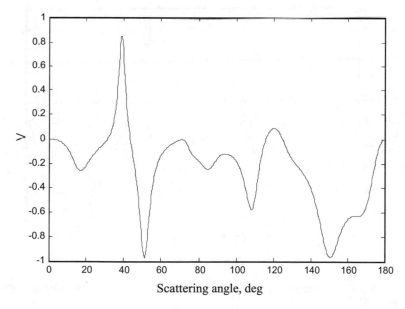

Fig. 3.15 Stokes parameters versus the scattering angle for the hemocyte suspension models at $d = 1.7 \cdot 10^{-7}$

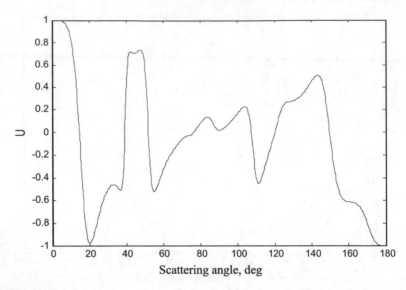

Fig. 3.16 Stokes parameters versus the scattering angle for the hemocyte suspension models at $d = 1.687 \cdot 10^{-7}$

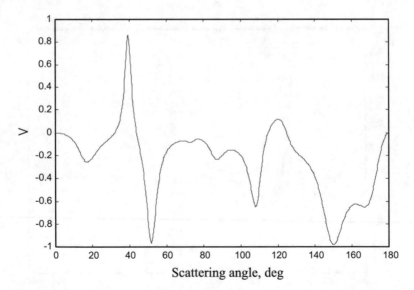

Fig. 3.17 Stokes parameters versus the scattering angle for the hemocyte suspension models at $d = 1.687 \cdot 10^{-7}$

with different structures using the same setup. Thus, it seems very likely to find a correlation between the electrophysical parameters of a biological substance and its biological properties using this approach.

References

1. A.N. Smirnov, *Diseases of the Blood*. A Series of Reference Books of a Practical Physician. M (2008)
2. M. Marlies, *Basics Hämatologie* (Elsevier GmbH, München, 2011)
3. K.G. Kulikov, A.I. Radin, Study of dispersion and absorption of an ensemble of spherical particles inside an optical cavity and new possibilities of predicting the optical characteristics of biological media by intracavity laser spectroscopy. Opt. Spectrosc. **82**(2), 228–236 (2002)
4. K.G. Kulikov, Investigation into the electrophysical characteristics of hemocytes by intracavity laser spectroscopy. ii: numerical simulation. Tech. Phys. **59**(5), 782–786 (2014)
5. Bohren, Huffman, *Absorption and Scattering of Light by Small Particles*. M (1986)
6. A. Aden, M. Kerker, Scattering of electromagnetic waves from two concentric spheres. J. Appl. Phys. **22**, 1242 (1951)
7. O.R. Cruzan, Translational addition theorems for spherical vector wave functions quart. Appl. Math. **20**(1), 33–40 (1962)
8. P.A, Bobbert, J. Vlieger, Light scattering by a sphere on a substrate. Physica A **137**(1), 209–241 (1986)
9. G. Videen, D. Ngo, P. Chylek, R.G. Pinnick, Light scattering from a sphere with an irregular inclusion. J. Opt. Soc. Am. **12**(5), 922–928 (1995)
10. M.Y. Balandin, E.P. Shurin, *Methods for Solving Linear Algebraic Large Dimension* (Novosibirsk, 2000)
11. H. Vorst van der, *Iterative Krylov Methods for Large Linear Sysmems* (Cambridge, 2003)
12. Y. Saad, *Iterative Methods for Sparse Linear Systems* (SIAM, 2003)
13. V.N. Kudashov, A.B. Plachenov, The natural oscillations in weakly inhomogeneous laser cavities. Opt. Spectrosc. **85**(2), 149–154 (1998)
14. V.N. Lopatin, A.V. Priezzhev, A.D. Aponasenko and others. *Methods Light Scattering in the Analysis of Dispersed Biological Media. M* (Fizmatlit, 2004)
15. G.V. Simonenko, V.V. Tuchin, *Optical Properties of Biological Tissues* (Saratov State University, Saratov, 2007) 48p
16. E. Libbert, *Fundamentals of General Biology. M* (Mir, 1982)

Chapter 4
Mathematical Models of the Interaction of Laser Radiation with Turbid Media

Abstract We construct the mathematical model for calculating the interaction of laser radiation with a turbid medium and the model for the prediction of the optical characteristics of blood (refractive index and absorption coefficient) and for the determination of the rate of blood flow in the capillary bed under irradiation of a laser beam is proposed.

4.1 Introduction

Now we proceed to the consideration of the principles of construction of mathematical models for calculating the interaction of laser radiation with a turbid medium. Turbid is called medium in which there is the absorption and scattering of radiation. One example of this is human skin tissue. As was noted in Chap. 1, the skin is a live multi-media containing various inclusions, such as, for example, blood vessels. Let's consider the basic periods of construction of the mathematical models describing the interaction of laser radiation with immunocompetent multilayered turbid media, such as human skin. (see Fig. 4.1). First we described the object of study. After that optical and physical parameters of all its components are defined. Next step is the calculation of the radiation in the environment, and (on some models) the calculation of the temperature fields. Distinctions between models become already appreciable at a period geometry construction. In most cases the skin is represented in the form of sequence of layers with various optical and thermophysical properties. The number of layers of the skin can vary from one to seven. The simple geometry includes only derma. The simplest geometry includes only the dermis [1]. This simplified model is used, for example, to simulate the treatment of acne laser light with a wavelength of 1450 nm.

The greatest number of layers of model is presented in [2]. Here, seven layers are located in skin: a cornual layer, epidermis, the upper derma, a derma with a superficial plexus of vessels, the lower derma, a derma with a deep plexus of vessels

© Springer International Publishing AG, part of Springer Nature 2018
K. Kulikov and T. Koshlan, *Laser Interaction with Heterogeneous Biological Tissue*, Biological and Medical Physics, Biomedical Engineering, https://doi.org/10.1007/978-3-319-94114-1_4

Fig. 4.1 The Scheme for the construction of models that describe the interaction laser light with objects

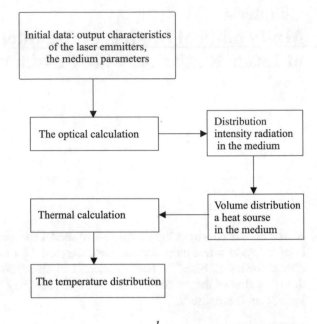

Fig. 4.2 Skin model. *1* the cornual layer, *2* epidermis, *3* the upper derma, *4* derma with a superficial plexus of vessels, *5* the lower derma, *6* derma with a deep plexus of vessels, *7* hypoderm

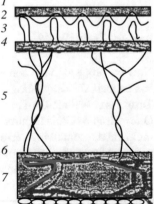

and a hypoderm (see Fig. 4.2). Each of layer has the optical characteristics. However such detailed separation of a skin into layers is used only for optical calculation.

Some authors identify the blood as a separate layer with the characteristics of pure blood, or as an object within the tissue. A single blood vessel is sometimes rectangular [3], or more cylindrical forms are usually placed in the dermis. The example of a model of the skin to the blood vessels of cylindrical form is shown in Fig. 4.3 [4].

In most cases the vessel wall has the same properties as a surrounding tissue. Models with vessels are located in the dermis, usually used to obtain the distribution

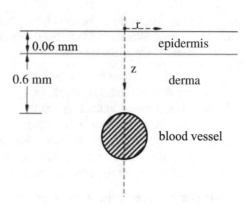

Fig. 4.3 Model of the skin with the blood vessels

of light intensity and temperature inside the veins. Optical properties are generally considered to be constant for a given wavelength and independent of temperature.

In article [5] it is assumed that the skin consists of the epidermis and dermis. Incident light first passes through the epidermis, where the largest coefficient has melanin, so the optical properties of the epidermis considered equal properties of melanin. The transmitted wave gets into the dermis, where it is mostly absorbed by hemoglobin, present in the surface layer of the dermis. The remaining radiation diffusely reflected from the collagen present in the rest of the dermis, and then passes through the layers of hemoglobin and melanin, partially absorbed. Such a description of the passage of light through the skin is used to calculate the coefficients of pigmentation and erythema.

In [3, 4] one describes another method for calculating the intensity distribution within the vessel. With the solution of the problem of electromagnetic diffraction on an infinite circular cylinder, the component of the electric field inside the cylindrical vessel is searched. These results let calculate the distribution function of heat sources inside the vessel.

4.2 An Electrodynamic Model of the Optical Characteristics of Blood and Capillary Blood Flow Rate

Application of lasers in biomedical investigations is based on the large variety of effects of interaction of light with biological objects. Optical methods are the most promising and are comparatively safe methods of study, being among the so-called noninvasive methods. However, the application of optical methods requires adequate theoretical models, whose development presents considerable difficulties.

It should be noted that a number of theoretical and experimental studies have been devoted to similar questions [6, 7]. In [6], the propagation of optical radiation through a biological medium (human skin) was modeled by the stochastic Monte Carlo method, which combines calculation schemes of real photon paths and the method of statistical weights. In [7], the absolute average flow rate of a biological fluid in microcapillaries and the flow direction were determined experimentally.

In this study, we will use an electrodynamic model, which makes it possible.

1. to vary such parameters of a biological structure as the real and imaginary parts of the refractive index of the blood, epidermis, upper dermis, and lower dermis;
2. to ascertain dependences between these parameters and the biological properties of blood irradiated by a laser beam;
3. to determine the rate of the blood flow in the capillary bed. This allows one to diagnose diseases whose manifestation is related with a decrease in the effective diameter of capillaries and with changes in the biophysical properties of blood.

Chapter is based on the result of the [8].

The biological tissue is represented in the form of layers with different optical characteristics (the epidermis, upper dermis, blood, and lower dermis), which are irradiated by a laser beam. The system of blood vessels is located in the upper layer of the dermis.

4.3   Reflection of a Plane Wave from a Layer with a Slowly Varying Thickness

In this section, we will find the coefficient of reflection from a layer with a slowly varying thickness.

Fig. 4.4 Schematic diagram of a model biological medium

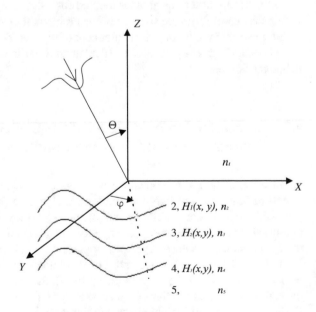

We will consider the optical system shown in Fig. 4.4. The system consists of four regions with different refractive indicesthe epidermis, the upper dermis, a blood vessel, and the lower dermis.

In order to attain the maximum possible correspondence between the structure of our model medium and that of the real object of study, we will represent the interfaces between the layers of the model medium in the form of curved surfaces

$$z_i = h_i(x, y), \; h_i(x, y) = c_i \sin(a_i x + b_i y). \tag{4.1}$$

In these expressions c_i, a_i, b_i are arbitrary constants obeying the conditions $a_i \ll 1, b_i \ll 1, c_i \ll 1, (i = \overline{1, 3})$.

Let a plane s- or p- polarized wave be incident on a layer at an angle θ.

$$E_{inc} = \exp(ik_{1x}x + ik_{1y}y - ik_{1z}z),$$

where

$$k_{1x} = kn_1 \sin(\theta) \sin(\phi), k_{1y} = kn_1 \sin(\theta) \cos(\phi),$$

$$k_{1z} = kn_1 \cos(\theta) \tag{4.2}$$

The reflected field must be found. We consider only the case of the p polarization. We will write the Maxwell equations for the jth layer of the medium,

$$\text{rot } \mathbf{E} = -i\omega\mu_0\mu_j \, \mathbf{H}, \text{rot } \mathbf{H} = i\omega\varepsilon_0\varepsilon_j \, \mathbf{E}, \text{div } \mathbf{E} = 0, \text{div } \mathbf{H} = 0 \tag{4.3}$$

Then, the electromagnetic field in the jth layer of the medium will satisfy the following wave equation

$$\Delta E + k^2 n_j^2 E = 0, \Delta H + k^2 n_j^2 H = 0, \tag{4.4}$$

where $k^2 = \omega^2 \varepsilon_0 \mu_0$, n_j is the complex refractive index of the jth layer ($j = \overline{1, 5}$), $n_j = n_j^o + i\chi_j$. We introduce the contracted coordinates

$$\xi_1 = \varepsilon x, \xi_2 = \varepsilon y, \xi_3 = \varepsilon z. \tag{4.5}$$

We will assume that the thicknesses of the layers H_1, H_2 and H_3 are slowly varying functions of the variables x and y. Let the ratio of the characteristic thickness of a layer to the characteristic linear size L be denoted as ε; then we obtain

$$H_1(x, y) = h_1(\xi_1, \xi_2)|_{\xi_1 = \varepsilon x, \xi_2 = \varepsilon y},$$

$$H_2(x, y) = h_2(\xi_1, \xi_2)|_{\xi_1 = \varepsilon x, \xi_2 = \varepsilon y},$$

$$H_3(x, y) = h_3(\xi_1, \xi_2)|_{\xi_1 = \varepsilon x, \xi_2 = \varepsilon y}.$$

The conditions that the tangential components of E and H should be continuous at the interfaces between media lead to the following boundary conditions

$$E_1|_{\xi_3=0} = E_2|_{\xi_3=0}, \; E_2|_{\xi_3=\varepsilon h_1(\xi_1,\xi_2)} = E_3|_{\xi_3=\varepsilon h_1(\xi_1,\xi_2)}, \tag{4.6}$$

$$E_3|_{\xi_3=\varepsilon h_2(\xi_1,\xi_2)} = E_4|_{\xi_3=\varepsilon h_2(\xi_1,\xi_2)},$$

$$E_4|_{\xi_3=\varepsilon h_3(\xi_1,\xi_2)} = E_5|_{\xi_3=\varepsilon h_3(\xi_1,\xi_2)}, \tag{4.7}$$

$$\frac{1}{n_1^2}\frac{\partial E_1}{\partial \xi_3}|_{\xi_3=0} = \frac{1}{n_2^2}\frac{\partial E_2}{\partial \xi_3}|_{\xi_3=0}, \tag{4.8}$$

$$\frac{1}{n_2^2}\left(\frac{\partial}{\partial \xi_3} - \varepsilon\frac{\partial h_1}{\partial \xi_1}\frac{\partial}{\partial \xi_1} - \varepsilon\frac{\partial h_1}{\partial \xi_2}\frac{\partial}{\partial \xi_2}\right)E_2|_{\xi_3=\varepsilon h_1(\xi_1,\xi_2)} =$$

$$= \frac{1}{n_3^2}\left(\frac{\partial}{\partial \xi_3} - \varepsilon\frac{\partial h_1}{\partial \xi_1}\frac{\partial}{\partial \xi_1} - \varepsilon\frac{\partial h_1}{\partial \xi_2}\frac{\partial}{\partial \xi_2}\right)E_3|_{\xi_3=\varepsilon h_1(\xi_1,\xi_2)}, \tag{4.9}$$

$$\frac{1}{n_3^2}\left(\frac{\partial}{\partial \xi_3} - \varepsilon\frac{\partial h_2}{\partial \xi_1}\frac{\partial}{\partial \xi_1} - \varepsilon\frac{\partial h_2}{\partial \xi_2}\frac{\partial}{\partial \xi_2}\right)E_3|_{\xi_3=\varepsilon h_2(\xi_1,\xi_2)} =$$

$$= \frac{1}{n_4^2}\left(\frac{\partial}{\partial \xi_3} - \varepsilon\frac{\partial h_2}{\partial \xi_1}\frac{\partial}{\partial \xi_1} - \varepsilon\frac{\partial h_2}{\partial \xi_2}\frac{\partial}{\partial \xi_2}\right)E_4|_{\xi_3=\varepsilon h_2(\xi_1,\xi_2)}, \tag{4.10}$$

$$\frac{1}{n_4^2}\left(\frac{\partial}{\partial \xi_3} - \varepsilon\frac{\partial h_3}{\partial \xi_1}\frac{\partial}{\partial \xi_1} - \varepsilon\frac{\partial h_3}{\partial \xi_2}\frac{\partial}{\partial \xi_2}\right)E_4|_{\xi_3=\varepsilon h_3(\xi_1,\xi_2)} =$$

$$= \frac{1}{n_5^2}\left(\frac{\partial}{\partial \xi_3} - \varepsilon\frac{\partial h_3}{\partial \xi_1}\frac{\partial}{\partial \xi_1} - \varepsilon\frac{\partial h_3}{\partial \xi_2}\frac{\partial}{\partial \xi_2}\right)E_5|_{\xi_3=\varepsilon h_3(\xi_1,\xi_2)}. \tag{4.11}$$

Since H_1, H_2, and H_3 are slowly varying functions of x and y, it is natural to seek the reflected field in the form of waves with slowly varying amplitudes and quickly oscillating phases:

$$E_1 = \exp\left(\frac{i}{\varepsilon}\tau_{inc}(\xi_1,\xi_2,\xi_3)\right) + \exp\left(\frac{i}{\varepsilon}\tau_{1ref}(\xi_1,\xi_2,\xi_3)\right) \times$$

$$\times A(\xi_1,\xi_2,\xi_3,\varepsilon_x,\varepsilon_y), \tag{4.12}$$

$$E_2 = \exp\left(\frac{i}{\varepsilon}\tau_{2elap}(\xi_1,\xi_2,\xi_3)\right)B^+(\xi_1,\xi_2,\xi_3,\varepsilon_x,\varepsilon_y)+$$

$$+ \exp\left(\frac{i}{\varepsilon}\tau_{3ref}(\xi_1,\xi_2,\xi_3)\right)B^-(\xi_1,\xi_2,\xi_3,\varepsilon_x,\varepsilon_y), \tag{4.13}$$

$$E_3 = \exp\left(\frac{i}{\varepsilon}\tau_{3elap}(\xi_1, \xi_2, \xi_3)\right) C^+(\xi_1, \xi_2, \xi_3, \varepsilon_x, \varepsilon_y) +$$

$$+ \exp\left(\frac{i}{\varepsilon}\tau_{4ref}(\xi_1, \xi_2, \xi_3)\right) C^-(\xi_1, \xi_2, \xi_3, \varepsilon_x, \varepsilon_y), \tag{4.14}$$

$$E_4 = \exp\left(\frac{i}{\varepsilon}\tau_{4elap}(\xi_1, \xi_2, \xi_3)\right) D^+(\xi_1, \xi_2, \xi_3, \varepsilon_x, \varepsilon_y) +$$

$$\exp\left(\frac{i}{\varepsilon}\tau_{5ref}(\xi_1, \xi_2, \xi_3)\right) D^-(\xi_1, \xi_2, \xi_3, \varepsilon_x, \varepsilon_y), \tag{4.15}$$

$$E_5 = \exp\left(\frac{i}{\varepsilon}\tau_{5elap}(\xi_1, \xi_2, \xi_3)\right) E(\xi_1, \xi_2, \xi_3, \varepsilon_x, \varepsilon_y). \tag{4.16}$$

Here, A, $B^\pm$, $C^\pm$, $D^\pm$ and E are the amplitudes, τ_{1ref}, τ_{2elap}, τ_{3ref}, τ_{3elap}, τ_{4ref}, τ_{5ref} and τ_{5elap} are unknown functions. By substituting the fields E_1, E_2, E_3, E_4 into (4.4), we obtain the equations for the amplitudes and eikonals:

$$\varepsilon^2 \Delta A + i\varepsilon(2\nabla A \nabla \tau_{1ref} + A\Delta\tau_{1ref}) + A(k^2 n_2^2 - \nabla\tau_{1ref}) = 0, \tag{4.17}$$

$$\varepsilon^2 \Delta B^+ + i\varepsilon(2\nabla B^+ \nabla \tau_{2elap} + B^+\Delta\tau_{2elap}) + B^+(k^2 n_2^2 - \nabla\tau_{2elap}) +$$

$$\varepsilon^2 \Delta B^- + i\varepsilon(2\nabla B^- \nabla \tau_{3ref} + B^-\Delta\tau_{3ref}) +$$

$$+ B^-(k^2 n_3^2 - \nabla\tau_{3ref}) = 0, \tag{4.18}$$

$$\varepsilon^2 \Delta C^+ + i\varepsilon(2\nabla C^+ \nabla \tau_{3elap} + C^+\Delta\tau_{3elap}) + C^+(k^2 n_3^2 - \nabla\tau_{3elap}) +$$

$$+ \varepsilon^2 \Delta C^- + i\varepsilon(2\nabla C^- \nabla \tau_{4ref} + C^-\Delta\tau_{4ref}) +$$

$$+ C^-(k^2 n_4^2 - \nabla\tau_{4ref}) = 0, \tag{4.19}$$

$$\varepsilon^2 \Delta D^+ + i\varepsilon(2\nabla D^+ \nabla \tau_{4elap} + D^+\Delta\tau_{4elap}) + D^+(k^2 n_4^2 - \nabla\tau_{4elap}) +$$

$$+ \varepsilon^2 \Delta D^- + i\varepsilon(2\nabla D^- \nabla \tau_{5ref} + D^-\Delta\tau_{5ref}) +$$

$$+ D^-(k^2 n_5^2 - \nabla\tau_{5ref}) = 0, \tag{4.20}$$

$$\varepsilon^2 \Delta E + i\varepsilon(2\nabla E \nabla \tau_{5elap} + E\Delta\tau_{5elap}) +$$

$$+ E(k^2 n_5^2 - \nabla\tau_{5elap}) = 0, \tag{4.21}$$

The amplitudes A, $B^{\pm}$, $C^{\pm}$, $D^{\pm}$ and E are sought in the form of series in powers of small parameters ε_x and ε_y.

$$A(\xi_1, \xi_2, \xi_3, \varepsilon_x, \varepsilon_y) = \sum_{i=0}^{\infty} \sum_{j=0}^{\infty} A_{ij}(\xi_1, \xi_2, \xi_3)(\varepsilon_x^i \cdot \varepsilon_y^j), \qquad (4.22)$$

$$B(\xi_1, \xi_2, \xi_3, \varepsilon_x, \varepsilon_y) = \sum_{i=0}^{\infty} \sum_{j=0}^{\infty} B_{ij}^{+}(\xi_1, \xi_2, \xi_3)(\varepsilon_x^i \cdot \varepsilon_y^j)+$$

$$+ \sum_{i=0}^{\infty} \sum_{j=0}^{\infty} B_{ij}^{-}(\xi_1, \xi_2, \xi_3)(\varepsilon_x^i \cdot \varepsilon_y^j), \qquad (4.23)$$

$$C(\xi_1, \xi_2, \xi_3, \varepsilon_x, \varepsilon_y) = \sum_{i=0}^{\infty} \sum_{j=0}^{\infty} C_{ij}^{+}(\xi_1, \xi_2, \xi_3)(\varepsilon_x^i \cdot \varepsilon_y^j)+$$

$$+ \sum_{i=0}^{\infty} \sum_{j=0}^{\infty} C_{ij}^{-}(\xi_1, \xi_2, \xi_3)(\varepsilon_x^i \cdot \varepsilon_y^j), \qquad (4.24)$$

$$D(\xi_1, \xi_2, \xi_3, \varepsilon_x, \varepsilon_y) = \sum_{i=0}^{\infty} \sum_{j=0}^{\infty} D_{ij}^{+}(\xi_1, \xi_2, \xi_3)(\varepsilon_x^i \cdot \varepsilon_y^j)+$$

$$+ \sum_{i=0}^{\infty} \sum_{j=0}^{\infty} D_{ij}^{-}(\xi_1, \xi_2, \xi_3)(\varepsilon_x^i \cdot \varepsilon_y^j), \qquad (4.25)$$

$$E(\xi_1, \xi_2, \xi_3, \varepsilon_x, \varepsilon_y) = \sum_{i=0}^{\infty} \sum_{j=0}^{\infty} E_{ij}(\xi_1, \xi_2, \xi_3)(\varepsilon_x^i \cdot \varepsilon_y^j). \qquad (4.26)$$

From (4.17)–(4.21) we obtain the equations for the eikonals

$$k^2 n_2^2 - \nabla \tau_{1ref} = 0, \quad k^2 n_2^2 - \nabla \tau_{2elap} = 0, \qquad (4.27)$$

$$k^2 n_3^2 - \nabla \tau_{3ref} = 0, \quad k^2 n_3^2 - \nabla \tau_{3elap} = 0, \qquad (4.28)$$

$$k^2 n_4^2 - \nabla \tau_{4ref} = 0, \quad k^2 n_4^2 - \nabla \tau_{4elap} = 0, \qquad (4.29)$$

$$k^2 n_5^2 - \nabla \tau_{5ref} = 0, \quad k^2 n_5^2 - \nabla \tau_{5elap} = 0. \qquad (4.30)$$

and amplitudes

$$\nabla A_{00} \nabla \tau_{1ref} = 0, \quad 2i \nabla A_{ij} \nabla \tau_{1ref} + \triangle A_{i-1,j-1} = 0, \tag{4.31}$$

$$\nabla B_{00}^+ \nabla \tau_{2elap} = 0, \quad 2i \nabla B_{ij}^+ \nabla \tau_{2elap} + \triangle B_{i-1,j-1}^+ = 0, \tag{4.32}$$

$$\nabla B_{00}^- \nabla \tau_{3ref} = 0, \quad 2i \nabla B_{ij}^- \nabla \tau_{3ref} + \triangle B_{i-1,j-1}^- = 0, \tag{4.33}$$

$$\nabla C_{00}^+ \nabla \tau_{3elap} = 0, \quad 2i \nabla C_{ij}^+ \nabla \tau_{3elap} + \triangle C_{i-1,j-1}^+ = 0, \tag{4.34}$$

$$\nabla C_{00}^- \nabla \tau_{4ref} = 0, \quad 2i \nabla C_{ij}^- \nabla \tau_{4ref} + \triangle C_{i-1,j-1}^- = 0, \tag{4.35}$$

$$\nabla D_{00}^+ \nabla \tau_{4elap} = 0, \quad 2i \nabla D_{ij}^+ \nabla \tau_{4elap} + \triangle D_{i-1,j-1}^+ = 0, \tag{4.36}$$

$$\nabla D_{00}^- \nabla \tau_{5ref} = 0, \quad 2i \nabla D_{ij}^- \nabla \tau_{5ref} + \triangle D_{i-1,j-1}^- = 0, \tag{4.37}$$

$$\nabla E_{00} \nabla \tau_{5elap} = 0, \quad 2i \nabla E_{ij} \nabla \tau_{5elap} + \triangle E_{i-1,j-1} = 0. \tag{4.38}$$

By solving (4.27)–(4.38) with regard to (4.22)–(4.26), we obtain the eikonals for the reflected and transmitted fields and the ray amplitudes A, B, C, D, and E (as the first approximation).

$$\tau_{1ref} = k_{2x}\xi_1 + k_{2y}\xi_2 + k_{2z}\xi_3, \quad k_{2x}^2 + k_{2y}^2 + k_{2z}^2 = k^2 n_2^2, \tag{4.39}$$

$$\tau_{2elap} = k_{2x}\xi_1 + k_{2y}\xi_2 - k_{2z}'\xi_3, \quad k_{2x}^2 + k_{2y}^2 + k_{2z}'^2 = k^2 n_2^2, \tag{4.40}$$

$$\tau_{3ref} = k_{3x}\xi_1 + k_{3y}\xi_2 + k_{3z}'\xi_3, \tag{4.41}$$

$$\tau_{3elap} = k_{3x}\xi_1 + k_{3y}\xi_2 - k_{3z}'\xi_3, \quad k_{3x}^2 + k_{3y}^2 + k_{3z}'^2 = k^2 n_3^2, \tag{4.42}$$

$$\tau_{4ref} = k_{4x}\xi_1 + k_{4y}\xi_2 + k_{4z}'\xi_3, \tag{4.43}$$

$$\tau_{4elap} = k_{4x}\xi_1 + k_{4y}\xi_2 - k_{4z}'\xi_3, k_{4x}^2 + k_{4y}^2 + k_{4z}'^2 = k^2 n_4^2, \tag{4.44}$$

$$\tau_{5ref} = k_{5x}\xi_1 + k_{5y}\xi_2 + k_{5z}'\xi_3, \quad k_{5x}^2 + k_{5y}^2 + k_{5z}'^2 = k^2 n_5^2, \tag{4.45}$$

$$\tau_{5elap} = k_{5x}\xi_1 + k_{5y}\xi_2 - k_{5z}\xi_3, \quad k_{5x}^2 + k_{5y}^2 + k_{5z}^2 = k^2 n_5^2, \tag{4.46}$$

where

$$k_{jx} = kn_j \sin(\theta)\sin(\phi), k_{jy} = kn_j \sin(\theta)\cos(\phi), k_{jz} = kn_j \cos(\theta), j = \overline{2,5}$$

$$A(\xi_1, \xi_2, \xi_3, \varepsilon_x, \varepsilon_y) = A_{\widehat{00}}(t_0) + \varepsilon_x \left[A_{\widehat{10}}(t_0) + \xi_3 A_{0000}(t_0) \right] +$$

$$+ \varepsilon_y \left[A_{\widehat{01}}(t_0) + \xi_3 A_{0000}(t_0) \right] +$$

$$+ \varepsilon_x \varepsilon_y \left[A_{\widehat{11}}(t_0) + \xi_3 A_{0000}(t_0) \right] + O(\varepsilon^2), \tag{4.47}$$

$$t_0 = \xi_1 + \xi_2 \frac{k_{2y}}{k_{2z}} - \xi_3 \frac{k_{2x}}{k_{2z}}, \qquad A_{0000}(t_0) = -\frac{1}{2i} \frac{\partial^2 A_{\widehat{00}}(t_0)}{\partial t_0^2} \frac{k^2 n_2^2}{k_{2z}^3},$$

$$B(\xi_1, \xi_2, \xi_3, \varepsilon_x, \varepsilon_y) = B_{\widehat{00}}^{+}(t_1) + \varepsilon_x \left[B_{\widehat{10}}^{+}(t_1) + B_{0000}^{+}(t_1)\xi_3 \right] +$$

$$+ \varepsilon_y \left[B_{\widehat{01}}^{+}(t_1) + B_{0000}^{+}(t_1)\xi_3 \right] + \varepsilon_x \varepsilon_y \left[B_{\widehat{11}}^{+}(t_1) + B_{0000}^{+}(t_1)\xi_3 \right] + B_{\widehat{00}}^{-}(t_2) +$$

$$+ \varepsilon_x \left[B_{\widehat{01}}^{-}(t_1) + B_{0000}^{-}(t_1)\xi_3 \right] + \varepsilon_y \left[B_{\widehat{01}}^{-}(t_2) + B_{0000}^{-}(t_2)\xi_3 \right]$$

$$+ \varepsilon_x \varepsilon_y \left[B_{\widehat{11}}^{-}(t_2) + B_{0000}^{-}(t_2)\xi_3 \right] + O(\varepsilon^2), \tag{4.48}$$

$$B_{0000}^{+}(t_1) = \frac{1}{2i} \frac{\partial^2}{\partial t_1^2} B_{\widehat{00}}^{+}(t_1) \frac{k^2 n_2^2}{k_{2z}^{\prime 3}}, \qquad B_{0000}^{-}(t_2) = -\frac{1}{2i} \frac{\partial^2}{\partial t_2^2} B_{\widehat{00}}^{-}(t_2) \frac{k^2 n_2^2}{k_{3z}^{\prime 3}},$$

$$t_1 = \xi_1 + \xi_2 \frac{k_{2y}}{k_{2z}'} + \xi_3 \frac{k_{2x}}{k_{2z}'}, \qquad t_2 = \xi_1 + \xi_2 \frac{k_{3y}}{k_{3z}'} - \xi_3 \frac{k_{3x}}{k_{3z}'},$$

$$C(\xi_1, \xi_2, \xi_3, \varepsilon_x, \varepsilon_y) = C_{\widehat{00}}^{+}(t_3) + \varepsilon_x \left[C_{\widehat{10}}^{+}(t_3) + C_{0000}^{+}(t_3)\xi_3 \right] +$$

$$\varepsilon_y \left[C_{\widehat{01}}^{+}(t_3) + C_{0000}^{+}(t_1)\xi_3 \right] + \varepsilon_x \varepsilon_y \left[C_{\widehat{11}}^{+}(t_3) + C_{0000}^{+}(t_1)\xi_3 \right] + C_{\widehat{00}}^{-}(t_4) +$$

$$+ \varepsilon_x \left[C_{\widehat{10}}^{-}(t_4) + C_{0000}^{-}(t_4)\xi_3 \right] + \varepsilon_y \left[C_{\widehat{01}}^{-}(t_2) + C_{0000}^{-}(t_4)\xi_3 \right] +$$

$$+ \varepsilon_x \varepsilon_y \left[C_{\widehat{11}}^{-}(t_4) + C_{0000}^{-}(t_4)\xi_3 \right] + O(\varepsilon^2), \tag{4.49}$$

$$C_{0000}^{+}(t_3) = \frac{1}{2i} \frac{\partial^2}{\partial t_3^2} C_{\widehat{00}}^{+}(t_3) \frac{k^2 n_2^2}{k_{3z}^{\prime 3}}, \qquad C_{0000}^{-}(t_4) = -\frac{1}{2i} \frac{\partial^2}{\partial t_4^2} C_{\widehat{00}}^{-}(t_4) \frac{k^2 n_2^2}{k_{4z}^{\prime 3}},$$

$$t_3 = \xi_1 + \xi_2 \frac{k_{3y}}{k_{3z}'} + \xi_3 \frac{k_{3x}}{k_{3z}'}, \qquad t_4 = \xi_1 + \xi_2 \frac{k_{4y}}{k_{4z}'} - \xi_3 \frac{k_{4x}}{k_{4z}'},$$

$$D(\xi_1, \xi_2, \xi_3, \varepsilon_x, \varepsilon_y) = D_{\widehat{00}}^{+}(t_5) + \varepsilon x \left[D_{\widehat{10}}^{+}(t_5) + D_{0000}^{+}(t_5)\xi_3 \right] +$$

$$+ \varepsilon y \left[D_{\widehat{01}}^{+}(t_5) + C_{0000}^{+}(t_5)\xi_3 \right] + \varepsilon_x \varepsilon_y \left[D_{\widehat{11}}^{+}(t_5) + D_{0000}^{+}(t_5)\xi_3 \right] + D_{\widehat{00}}^{-}(t_6) +$$

$$+ \varepsilon_x \left[D_{\widehat{10}}^{-}(t_6) + D_{0000}^{-}(t_6)\xi_3 \right] + \varepsilon_y \left[D_{\widehat{01}}^{-}(t_6) + D_{0000}^{-}(t_6)\xi_3 \right] +$$

$$+ \varepsilon_x \varepsilon_y \left[D_{\widehat{11}}^{-}(t_6) + D_{0000}^{-}(t_6)\xi_3 \right] + O(\varepsilon^2), \tag{4.50}$$

$$D_{0000}^{+}(t_5) = \frac{1}{2i} \frac{\partial^2}{\partial t_5^2} D_{\widehat{00}}^{+}(t_5) \frac{k^2 n_2^2}{k_{4z}^{\prime 3}}, \qquad D_{0000}^{-}(t_6) = -\frac{1}{2i} \frac{\partial^2}{\partial t_6^2} D_{\widehat{00}}^{-}(t_6) \frac{k^2 n_2^2}{k_{5z}^{\prime 3}},$$

$$t_5 = \xi_1 + \xi_2 \frac{k_{4y}}{k'_{4z}} + \xi_3 \frac{k_{4x}}{k'_{4z}}, \quad t_6 = \xi_1 + \xi_2 \frac{k_{5y}}{k'_{5z}} - \xi_3 \frac{k_{5x}}{k'_{5z}}$$

$$E(\xi_1, \xi_2, \xi_3, \varepsilon_x, \varepsilon_y) = E_{\widehat{00}}(t_7) + \varepsilon_x \left[E_{\widehat{10}}(t_7) + \xi_3 E_{0000}(t_7) \right] + \tag{4.51}$$

$$+\varepsilon_y \left[E_{\widehat{01}}(t_7) + \xi_3 E_{0000}(t_7) \right] + \varepsilon_x \varepsilon_y \left[E_{\widehat{11}}(t_7) + \xi_3 E_{0000}(t_7) \right] + O(\varepsilon^2),$$

$$E_{0000}(t_7) = \frac{1}{2i} \frac{\partial^2 E_{\widehat{00}}(t_7)}{\partial t_7^2} \frac{k^2 n_2^2}{k'^3_{5z}}, \quad t_7 = \xi_1 + \xi_2 \frac{k_{5y}}{k_{5z}} + \xi_3 \frac{k_{5x}}{k_{5z}}.$$

The substitution of (4.12)–(4.16) into (4.6)–(4.11) with regard to (4.39)–(4.46) and (4.47)–(4.51) gives rise to a recurrent system of equations for the successive determination of terms of series (4.22)–(4.26).

$$1 + A_{\widehat{00}} = B_{\widehat{00}}^+ + B_{\widehat{00}}^-,$$

$$B_{\widehat{00}}^+ \exp(-h_1 i k'_{2z}) + B_{\widehat{00}}^- \exp(h_1 i k'_{3z}) = C_{\widehat{00}}^+ \exp(-h_1 i k'_{3z}) + C_{\widehat{00}}^- \exp(h_1 i k'_{4z}),$$

$$C_{\widehat{00}}^+ \exp(-h_2 i k'_{3z}) + C_{\widehat{00}}^- \exp(h_2 i k'_{4z}) = D_{\widehat{00}}^+ \exp(-h_2 i k'_{4z}) + D_{\widehat{00}}^- \exp(h_2 i k'_{5z})$$

$$D_{\widehat{00}+} \exp(-h_3 i k'_{4z}) + D_{\widehat{00}}^- \exp(h_3 i k'_{5z}) = E_{\widehat{00}} \exp(-h_3 i k_{5z}),$$

$$\frac{1}{n_1^2} \left(-i k_{1z} + i k_{2z} A_{\widehat{00}} \right) = \frac{1}{n_2^2} \left(-i k'_{2z} B_{\widehat{00}}^+ + i k'_{3z} B_{\widehat{00}}^- \right),$$

$$\frac{1}{n_2^2} \left(-i k'_{2z} B_{\widehat{00}}^+ \exp(-h_1 i k'_{2z}) + i k'_{3z} B_{\widehat{00}}^- \exp(h_1 i k'_{3z}) \right) =$$

$$= \frac{1}{n_3^2} \left(-i k'_{3z} C_{\widehat{00}}^+ \exp(-h_1 i k'_{3z}) + i k'_{4z} C_{\widehat{00}}^- \exp(h_1 i k'_{4z}) \right),$$

$$\frac{1}{n_3^2} \left(-i k'_{3z} C_{\widehat{00}}^+ \exp(-h_2 i k'_{3z}) + i k'_{4z} C_{\widehat{00}}^- \exp(h_2 i k'_{4z}) \right) =$$

$$= \frac{1}{n_4^2} \left(-i k'_{4z} D_{\widehat{00}}^+ \exp(-h_2 i k'_{4z}) + i k'_{5z} D_{\widehat{00}}^- \exp(h_2 i k'_{5z}) \right),$$

$$\frac{1}{n_4^2} \left(-i k'_{4z} D_{\widehat{00}}^+ \exp(-h_2 i k'_{4z}) + i k'_{5z} D_{\widehat{00}}^- \exp(h_2 i k'_{5z}) \right) =$$

$$= \frac{1}{n_5^2} \left(-i k_{5z} E_{\widehat{00}} \exp(-h_3 i k_{5z}) \right).$$

From this system, one can find the reflection coefficient in the principal approximation for the reflected field.

Now, we will pass to the derivation of the formulas for the reflection of a Gaussian beam. This problem will be solved by expansion of counter propagating waves in terms of plane waves in the region of medium 1, their reflection by layer 2, and reverse transformation with a subsequent Huygens–Fresnel integral transformation to obtain the field in the initial section.

4.4   Reflection of a Gaussian Beam from a Layer with a Slowly Varying Thickness

Let a Gaussian beam with an arbitrary transverse field distribution be incident on a layer at an angle θ. We will relate the coordinate system (x', y', z') to the direction of incidence of the beam. The reflected field will be sought in the coordinate system (x'', y'', z''). Let the incident field have the form along the straight line $z' = 0$

$$E_{inc}|_{z'=0} = \Phi(\xi_1', \xi_2')|_{\xi_1'=\varepsilon x', \xi_2'=\varepsilon y'}.$$

Let the function Φ runs sufficiently rapidly to zero starting from the distances of the order of $O(1/\varepsilon)$-axis of z'. We will write the identity:

$$\Phi(\xi_1', \xi_2') = \frac{1}{(2\pi)^2} \int_{-\infty}^{\infty}\int_{-\infty}^{\infty} \exp[ik_{1x}^{\wedge}\xi_1' + ik_{1y}^{\wedge}\xi_2']dk_{1x}^{\wedge}dk_{1y}^{\wedge} \int_{-\infty}^{\infty}\int_{-\infty}^{\infty} \times$$

$$\times \exp[-ik_{1x}^{\wedge}\xi_1^{\wedge} - ik_{1y}^{\wedge}\xi_2^{\wedge}]\Phi(\xi_1^{\wedge}, \xi_2^{\wedge})d\xi_1^{\wedge}d\xi_2^{\wedge}.$$

Then, the incident field can be represented as

$$E_{inc} = \frac{1}{(2\pi)^2} \int_{-\infty}^{\infty}\int_{-\infty}^{\infty} dk_{1x}^{\wedge}dk_{1y}^{\wedge} \times$$

$$\times \exp[-iz'\sqrt{k^2 n_1^2 - \varepsilon^2 x^2 k_{1x}^{2\wedge} - \varepsilon^2 y^2 k_{1y}^{2\wedge}} + ik_{1x}^{\wedge}\xi_1' + ik_{1y}^{\wedge}\xi_2'] \times$$

$$\times \int_{-\infty}^{\infty}\int_{-\infty}^{\infty} d\xi_1^{\wedge}d\xi_2^{\wedge} \exp[-ik_{1x}^{\wedge}\xi_1^{\wedge} - ik_{1y}^{\wedge}\xi_2^{\wedge}]\Phi(\xi_1^{\wedge}, \xi_2^{\wedge})$$

We note that the incident field satisfies the Helmholtz equation. By expanding the exponent of the exponential into a series in terms of a small parameter, we obtain the following expression for the field:

$$E_{inc} = \frac{1}{(2\pi)^2} \int_{-\infty}^{\infty}\int_{-\infty}^{\infty} dk_{1x}^{\wedge}dk_{1y}^{\wedge} \times$$

$$\times \exp\left[-iz'\left(1 - \frac{0.5\varepsilon^2 y^2 k_{1y}^{2\wedge}}{n_1^2 k^2} - \frac{0.5\varepsilon^2 x^2 k_{1x}^{2\wedge}}{n_1^2 k^2} - \frac{\varepsilon^2 x^2 \varepsilon^2 y^2 k_{1x}^{2\wedge} k_{1y}^{2\wedge}}{4n_1^4 k^4} + O(\varepsilon^4)\right)\right] \times$$

$$\times \exp[ik_{1x}^{\wedge}\varepsilon x' + ik_{1y}^{\wedge}\varepsilon y'] \int_{-\infty}^{\infty}\int_{-\infty}^{\infty} d\xi_1^{\wedge} d\xi_2^{\wedge} \exp[-ik_{1x}^{\wedge}\xi_1^{\wedge} - ik_{1y}^{\wedge}\xi_2^{\wedge}]\Phi(\xi_1^{\wedge}, \xi_2^{\wedge}).$$

If $|k_{1x}^{\wedge}| \ll k$, $|k_{1y}^{\wedge}| \ll k$, the square root in the exponential

$$\sqrt{k^2 n_1^2 - \varepsilon^2 x^2 k_{1x}^{2\wedge} - \varepsilon^2 y^2 k_{1y}^{2\wedge}}$$

can be expanded into a series in which only terms quadratic in k_{1x} and k_{1y} would be retained. Then,

$$E_{inc} = \frac{1}{(2\pi)^2}\int_{-\infty}^{\infty}\int_{-\infty}^{\infty} dk_{1x}^{\wedge} dk_{1y}^{\wedge} \times$$

$$\times \exp\left[-iz'\left(1 - \frac{0.5\varepsilon^2 y^2 k_{1y}^{2\wedge}}{n_1^2 k^2} - \frac{0.5\varepsilon^2 x^2 k_{1x}^{2\wedge}}{n_1^2 k^2} + O(\varepsilon^4)\right) + ik_{1x}^{\wedge}\varepsilon x' + ik_{1y}^{\wedge}\varepsilon y'\right] \times$$

$$\times \int_{-\infty}^{\infty}\int_{-\infty}^{\infty} d\xi_1^{\wedge} d\xi_2^{\wedge} \exp[-ik_{1x}^{\wedge}\xi_1^{\wedge} - ik_{1y}^{\wedge}\xi_2^{\wedge}]\Phi(\xi_1^{\wedge}, \xi_2^{\wedge}).$$

Let us relate the coordinate systems (x', y', z') and (x, y, z)

$$kn_1 x' = k_{11}x + k_{12}y + k_{13}z,$$

$$kn_1 y' = k_{21}x + k_{22}y + k_{23}z,$$

$$kn_1 z' = k_{31}x + k_{32}y + k_{33}z,$$

$$k_{11} = kn_1 a_{11}, k_{12} = kn_1 a_{12}, k_{13} = kn_1 a_{13}, k_{21} = kn_1 a_{21}, \tag{4.52}$$

$$k_{22} = kn_1 a_{22}, k_{23} = kn_1 a_{23}, k_{31} = kn_1 a_{31}, k_{32} = kn_1 a_{32}, k_{33} = kn_1 a_{33}, \tag{4.53}$$

$$a_{11} = \cos(\varphi)\cos(\psi) - \sin(\varphi)\cos(\theta)\sin(\psi), \tag{4.54}$$

$$a_{12} = -\sin(\varphi)\cos(\psi) - \cos(\varphi)\cos(\theta)\sin(\psi), \tag{4.55}$$

$$a_{13} = \sin(\theta)\sin(\psi), \tag{4.56}$$

$$a_{21} = \cos(\varphi)\sin(\psi) + \sin(\varphi)\cos(\theta)\cos(\psi), \tag{4.57}$$

$$a_{22} = -\sin(\varphi)\sin(\psi) + \cos(\varphi)\cos(\theta)\cos(\psi), \tag{4.58}$$

$$a_{33} = \cos(\theta), \, a_{23} = -\sin(\theta)\cos(\psi), \, a_{31} = \sin(\varphi)\sin(\theta), \qquad (4.59)$$

$$a_{32} = \cos(\varphi)\sin(\theta). \qquad (4.60)$$

In the coordinate system (x, y, z), the incident field is written as

$$E_{inc} = \frac{1}{(2\pi)^2} \int_{-\infty}^{\infty} \int_{-\infty}^{\infty} dk_{1x}^{\wedge} dk_{1y}^{\wedge} \times$$

$$\times \exp[i(xk_{1x} + yk_{1y} - zk_{1z}) \int_{-\infty}^{\infty} \int_{-\infty}^{\infty} d\xi_1^{\wedge} d\xi_2^{\wedge} \exp[-ik_{1x}^{\wedge}\xi_1^{\wedge} - ik_{1y}^{\wedge}\xi_2^{\wedge}]\Phi(\xi_1^{\wedge}, \xi_2^{\wedge}),$$

where

$$k_{1x} = -\left[k_{31}\left(1 - \frac{\varepsilon^2 y^2 k_{1y}^{2\wedge}}{2k^2 n_1^2} - \frac{\varepsilon^2 x^2 k_{1x}^{2\wedge}}{2k^2 n_1^2} \right) + \varepsilon \frac{k_{1x}^{\wedge}}{kn_1} k_{11} - \varepsilon \frac{k_{1y}^{\wedge}}{kn_1} k_{21} \right] +$$

$$+ O(\varepsilon^4), \qquad (4.61)$$

$$k_{1y} = -\left[k_{32}\left(1 - \frac{\varepsilon^2 y^2 k_{1y}^{2\wedge}}{2k^2 n_1^2} - \frac{\varepsilon^2 x^2 k_{1x}^{2\wedge}}{2k^2 n_1^2} \right) + \varepsilon \frac{k_{1x}^{\wedge}}{kn_1} k_{12} - \varepsilon \frac{k_{1y}^{\wedge}}{kn_1} k_{22} \right] +$$

$$+ O(\varepsilon^4), \qquad (4.62)$$

$$k_{1z} = \left[k_{33}\left(1 - \frac{\varepsilon^2 y^2 k_{1y}^{2\wedge}}{2k^2 n_1^2} - \frac{\varepsilon^2 x^2 k_{1x}^{2\wedge}}{2k^2 n_1^2} \right) + \varepsilon \frac{k_{1x}^{\wedge}}{kn_1} k_{13} - \varepsilon \frac{k_{1y}^{\wedge}}{kn_1} k_{23} \right] +$$

$$+ O(\varepsilon^4). \qquad (4.63)$$

4.5   The Reflected Field

Upon reflection, each spectral component $\exp[ixk_{1x} + iyk_{1y} - izk_{1z}]$ gives rise to a reflected wave $A(\xi_1, \xi_2, \xi_3, k_{1x}, k_{1y})\exp[ixk_{1x} + iyk_{1y} + izk_{1z}]$, where A is the amplitude determined by formula (4.47) and k_{1x}, k_{1y} and k_{1z} are given by formulas (4.61)–(4.63). The reflected field will be written as

$$E_{ref} = \frac{1}{(2\pi)^2} \int_{-\infty}^{\infty} \int_{-\infty}^{\infty} dk_{1x}^{\wedge} dk_{1y}^{\wedge} \exp(ixk_{1x} + iyk_{1y} + izk_{1z})$$

$$A(\xi_1, \xi_2, \xi_3, k_{1x}, k_{1y}) \int_{-\infty}^{\infty} \int_{-\infty}^{\infty} d\xi_1^{\wedge} d\xi_2^{\wedge} \exp(-ik_{1x}^{\wedge}\xi_1^{\wedge} - ik_{1y}^{\wedge}\xi_2^{\wedge})\Phi(\xi_1^{\wedge}, \xi_2^{\wedge}).$$

We note that reflected field also satisfies the Helmholtz equation. Let us pass to the coordinate system (x'', y'', z''), related to the reflected field:

$$kn_1x = k_{11}x'' + k_{21}y'' + k_{31}z'',$$

$$kn_1y = k_{12}x'' + k_{22}y'' + k_{32}z'',$$

$$kn_1z = k_{13}x'' + k_{23}y'' + k_{33}z'',$$

where $k_{11}, k_{12}, k_{13}, k_{21}, k_{22}, k_{23}, k_{31}, k_{32}, k_{33}$ are determined by relations (4.52)–(4.60). In this coordinate system, the reflected field takes the form

$$E_{ref} = \exp(-i(k + O(\varepsilon^4))z'')\frac{1}{(2\pi)^2}\int_{-\infty}^{\infty}\int_{-\infty}^{\infty}dk_{1x}^{\wedge}dk_{1y}^{\wedge}\times$$

$$\times \exp\left[-z''i\left(-\frac{\varepsilon^2 y^2 k_{1y}^{\wedge}}{2k^2 n_1^2} - \frac{\varepsilon^2 x^2 k_{1x}^{\wedge}}{2k^2 n_1^2} + O(\varepsilon^3)\right)\right]\times$$

$$\times \exp[ik_{1x}^{\wedge}\xi_1''(a_{11}^2 + a_{22}a_{12} + a_{13}^2) + ik_{1y}^{\wedge}\xi_1''(a_{21}a_{11} + a_{22}^2 + a_{23}a_{13})+$$

$$+ik_{1x}^{\wedge}\xi_2''(a_{11}a_{21} + a_{22}^2 a_{13}a_{23}) + ik_{1y}^{\wedge}\xi_2''(a_{21}^2 + a_{12}a_{22} + a_{23}^2)]A(\xi_1, \xi_2, \xi_3, k_{1x}, k_{1y})\times$$

$$\times \int_{-\infty}^{\infty}\int_{-\infty}^{\infty}d\xi_1^{\wedge}d\xi_2^{\wedge}\exp(\ ik_{1x}^{\wedge}\xi_1^{\wedge}\quad ik_{1y}^{\wedge}\xi_2^{\wedge})\Phi(\zeta_1^{\wedge}, \zeta_2^{\wedge}). \qquad (4.64)$$

In order to calculate the field of the reflected beam, one needs to reexpand the amplitude A in terms of a small parameter, substitute this expansion into the formula for the reflected field, and perform the integration. In the coordinate system (x'', y'', z''), the reflected field in the section $z'' = 0$ related to the ray takes the form

$$E_{ref} = \frac{1}{(2\pi)^2}\int_{-\infty}^{\infty}\int_{-\infty}^{\infty}dk_{1x}^{\wedge}dk_{1y}^{\wedge}\exp[ik_{1x}^{\wedge}\xi_1''(a_{11}^2 + a_{22}a_{12} + a_{13}^2)+$$

$$+ik_{1y}^{\wedge}\xi_1''(a_{21}a_{11} + a_{22}^2 + a_{23}a_{13})+$$

$$+ik_{1x}^{\wedge}\xi_2''(a_{11}a_{21} + a_{22}^2 + a_{13}a_{23}) + ik_{1y}^{\wedge}\xi_2''(a_{21}^2 + a_{12}a_{22} + a_{23}^2)]\times$$

$$\times A(\xi_1, \xi_2, \xi_3, k_{1x}, k_{1y})\int_{-\infty}^{\infty}\int_{-\infty}^{\infty}d\xi_1^{\wedge}d\xi_2^{\wedge}\exp(-ik_{1x}^{\wedge}\xi_1^{\wedge} - ik_{1y}^{\wedge}\xi_2^{\wedge})\times$$

$$\times \Phi(\xi_1^{\wedge}, \xi_2^{\wedge}) + O(\varepsilon^2) \qquad (4.65)$$

By expanding A in terms of a small parameter ε at $z'' = 0$, we have

$$A\left(\xi_1''\frac{k_x^1}{kn_1}+\xi_2''\frac{k_y^1}{kn_1}-\left(\xi_1''\frac{k_{13}}{kn_1}\frac{k_{1x}k_{1y}}{k_{1z}}+\xi_2''\frac{k_{23}}{kn_1}\frac{k_{1x}k_{1y}}{k_{1z}}\right),k_{1y},k_{1x}\right)=$$

$$= A_{\widehat{oo}}(\xi_1''^{\sim}+\xi_2''^{\sim},k_{1y},k_{1x})-$$

$$-\varepsilon_x\left[A_{\widehat{10}}(\xi_1''^{\sim}+\xi_2''^{\sim},k_{1y},k_{1x})+\frac{k_{13}}{kn_1}\xi_1''A_{0000}(\xi_1''^{\sim}+\xi_2''^{\sim},k_{1y},k_{1x})\right]-$$

$$-\varepsilon_y\left[A_{\widehat{01}}(\xi_1''^{\sim}+\xi_2''^{\sim},k_{1y},k_{1x})+\frac{k_{23}}{kn_1}\xi_2''A_{0000}(\xi_1''^{\sim}+\xi_2''^{\sim},k_{1y},k_{1x})\right]-$$

$$-\varepsilon_x\varepsilon_y\left[A_{\widehat{11}}(\xi_1''^{\sim}+\xi_2''^{\sim},k_{1y},k_{1x})+\frac{k_{13}}{kn_1}\xi_1''A_{0000}(\xi_1''^{\sim}+\xi_2''^{\sim},k_{1y},k_{1x})\right]-$$

$$-\varepsilon_x\varepsilon_y\left[\frac{k_{23}}{kn_1}\xi_2''A_{0000}(\xi_1''^{\sim}+\xi_2''^{\sim},k_{1y},k_{1x})\right]-$$

$$-\frac{\partial A_{0000}(\xi_1''^{\sim}+\xi_2''^{\sim},k_{1y},k_{1x})}{\partial k_{1x}}\frac{(k_{1x}^{\wedge}k_{11}-k_{1y}^{\wedge}k_{21})}{kn_1}-$$

$$-\frac{\partial A_{0000}(\xi_1''^{\sim}+\xi_2''^{\sim},k_{1y},k_{1x})}{\partial k_{1y}}\frac{(k_{1x}^{\wedge}k_{12}-k_{1y}^{\wedge}k_{22})}{kn_1}+O(\varepsilon^2). \qquad (4.66)$$

The substitution of (4.66) into (4.65) and integration yield the following expression for the reflected field along the beam axis $z''=0$:

$$E_{ref}=\frac{A_{\widehat{oo}}(\xi_1''^{\sim}+\xi_2''^{\sim},k_{1y},k_{1x})\Phi(\xi_1'',\xi_2'')}{\alpha}- \qquad (4.67)$$

$$-\frac{\varepsilon_x}{\alpha}\left[A_{\widehat{10}}(\xi_1''^{\sim}+\xi_2''^{\sim},k_{1y},k_{1x})+\frac{k_{13}}{kn_1}\xi_1''A_{0000}(\xi_1''^{\sim}+\xi_2''^{\sim},k_{1y},k_{1x})\right]\Phi(\xi_1'',\xi_2'')-$$

$$-\frac{\varepsilon_y}{\alpha}\left[A_{\widehat{01}}(\xi_1''^{\sim}+\xi_2''^{\sim},k_{1y},k_{1x})+\frac{k_{23}}{kn_1}\xi_2''A_{0000}(\xi_1''^{\sim}+\xi_2''^{\sim},k_{1y},k_{1x})\right]\Phi(\xi_1'',\xi_2'')-$$

$$-\frac{\varepsilon_x\varepsilon_y}{\alpha}\left[A_{\widehat{11}}(\xi_1''^{\sim}+\xi_2''^{\sim},k_{1y},k_{1x})+\frac{k_{13}}{kn_1}\xi_1''A_{0000}(\xi_1''^{\sim}+\xi_2''^{\sim},k_{1y},k_{1x})\right]\Phi(\xi_1'',\xi_2'')-$$

$$-\frac{\varepsilon_x\varepsilon_y}{\alpha}\left[\frac{k_{23}}{kn_1}\xi_2''A_{0000}(\xi_1''^{\sim}+\xi_2''^{\sim},k_{1y},k_{1x})\right]\Phi(\xi_1'',\xi_2'')-$$

$$-\left\{\frac{\varepsilon_x k_x^o}{ikn_1\alpha}\left[\frac{\partial A_{\widehat{oo}}(\xi_1''^{\sim}+\xi_2''^{\sim},k_{1y},k_{1x})}{\partial k_{1x}}+\frac{\partial A_{\widehat{oo}}(\xi_1''^{\sim}+\xi_2''^{\sim},k_{1y},k_{1x})}{\partial k_{1y}}\right]\frac{\partial\Phi(\xi_1'',\xi_2'')}{\partial\xi_1''}\right\}-$$

$$-\left\{\frac{\varepsilon_y k_y^o}{i k n_1 \alpha}\left[\frac{\partial A_{\widehat{oo}}(\xi_1''^{\sim}+\xi_2''^{\sim},k_{1y},k_{1x})}{\partial k_{1x}}+\frac{\partial A_{\widehat{oo}}(\xi_1''^{\sim}+\xi_2''^{\sim},k_{1y},k_{1x})}{\partial k_{1y}}\right]\frac{\partial \Phi(\xi_1'',\xi_2'')}{\partial \xi_2''}\right\}+$$

$$+O(\varepsilon^2).$$

Note that the reflected field depends on the parameters of the incident beam (the angle of incidence, the field distribution in a fixed section), the geometry of the boundaries of the reflecting medium, the refractive index. The reflected field is represented as the sum of the principal and the correction terms of the asymptotic of the small parameter with precision $O(\varepsilon^2)$.

For fixed parameters of the system are two main factors that determine the distortion of the field of the incident beam upon reflection. The first factor, call it geometric, is described by the term in square brackets in (4.67). The reflected field is obtained by multiplying the incident beam field to the local reflection coefficient of a plane wave of unit amplitude incident on the medium at the same angle as the beam.

The second factor is called it the diffusion. It described by the term in the curly brackets of (4.67). This term describes the distortion of the beam reflected by the transverse to the direction of propagation of the reflected beam diffusion amplitude. It should be noted that the reflection formulas were obtained for the field of a beam with an arbitrary transverse distribution incident at an arbitrary angle to a certain surface of a body with an arbitrary refractive index for the s- and p- polarization of the incident beam. The results are represented as an asymptotic form with a small parameter having the meaning of the ratio of the characteristic scale of variation of the profile of the boundary of the body to the characteristic distance over which this variation took place. The calculations were performed with an error of the order of the quadratic terms of the asymptotic form. The resultant formulas are finite for any values of the system parameters except the angle of incidence of the beam.

The formulas are nonuniform on the angle of incidence. Upon an increase in the angle of incidence, the correction terms of the asymptotic form will also increase, which shows a growing distortion of the beam. When the angle of incidence is equal $90°$ then the wave beam is completely destroyed. Thus, the reflection formulas are valid in the range $0°-89°$.

The expressions for the reflected H. field are derived in a similar manner. In what follows, we consider the reflected field in the main approximation.

4.6   Calculation of the Rate of Blood Flow in a Capillary

In order to calculate the rate of the blood flow in a capillary, we will use the Galilean transformation. For definiteness, the blood vessel is assumed to be oriented along the Ox axis. Then

$$x = x' + v_x t, \; y = y'. \tag{4.68}$$

We will substitute the formula (4.68) into (4.69) and expand the latter expression into Taylor series in terms of υ_x, retaining only linear terms. The substitution of this expansion into (4.69) yields the dependence of the intensity on the rate of blood flow in the capillary at the time instant t. The intensity of radiation is determined as

$$I = |E_\perp|^2 + |E_\parallel|^2, \tag{4.69}$$

$$E_\perp = \cos(\theta)E_z + \sin(\theta)E_x,$$

$$E_\parallel = \sin(\theta)E_z - \cos(\theta)E_x,$$

where E_x and E_z are given by the following expressions

$$\frac{\partial E_z}{\partial y} - \frac{\partial E_y}{\partial z} = -i\omega\mu_0\mu_j H_x, \quad \frac{\partial E_x}{\partial z} - \frac{\partial E_z}{\partial x} = -i\omega\mu_0\mu_j H_y, \tag{4.70}$$

$$\frac{\partial E_y}{\partial x} - \frac{\partial E_x}{\partial y} = -i\omega\mu_0\mu_j H_z, \quad \frac{\partial H_z}{\partial y} - \frac{\partial H_y}{\partial z} = i\omega\varepsilon_0\varepsilon_j E_x, \tag{4.71}$$

$$\frac{\partial H_x}{\partial z} - \frac{\partial H_z}{\partial x} = i\omega\varepsilon_0\varepsilon_j E_y, \quad \frac{\partial H_y}{\partial x} - \frac{\partial H_x}{\partial y} = i\omega\varepsilon_0\varepsilon_j E_z. \tag{4.72}$$

Formulas (4.70)–(4.72) correspond to the system of the Maxwell equations (4.3) in a Cartesian coordinate system. Thus, we obtained formulas allowing one to determine the explicit dependence of the intensity of laser radiation as a function of the refractive index and absorption coefficient for the system of blood vessels located in the upper dermis on the rate of blood flow in the capillary bed at the time instant t and on the coordinate system.

The following investigation and the analysis of the dependences presented will be performed by numerical methods.

4.7   Numerical Calculations for a Model Medium and Conclusions

Let us consider the model medium shown in Fig. 4.4. The parameters of the medium are as follows. The refractive indices of the layers are equal to $n_2^o = 1.50, n_3^o = 1.40,$ $n_4^o = 1.35, n_5^o = 1.40$; the characteristic thicknesses of the layers are $d_2 = 65 \cdot 10^{-6},$ $d_3 = 565 \cdot 10^{-6}, d_4 = 90 \cdot 10^{-6}, n_1^o = 1, \chi_1 = 0, \chi_2 = \chi_3 = \chi_4 = \chi_5 = 10^{-5}, a_1 = -0.0024, b_1 = 0.020, a_2 = 0.021, b_2 = 0.030, a_3 = 0.041, b_3 = 0.051, c_1 = c_2 = c_3 = 10^{-2}$ and the wavelength is $\lambda = 0.63\,\mu\text{m}$ (the radiation wavelength of a He−Ne laser).

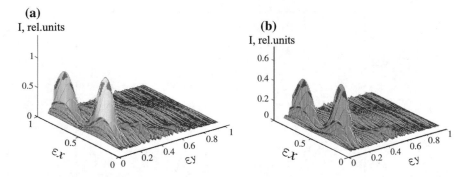

Fig. 4.5 Dependences of the intensity of radiation of a HeNe laser at a wavelength of $0.63\,\mu m$ on (**a, b**), $\theta = 0°$, $\varphi = 0°$, $\psi = 0°$, $\chi = 10^{-5}$ (absorption coefficient of blood) and refractive indices of blood $n_4 = 1.35$ (**a**) and $n_4 = 1.35003$ (**b**)

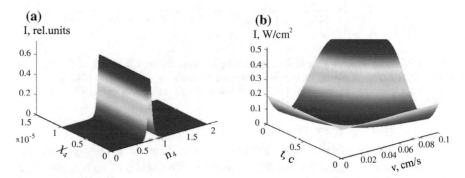

Fig. 4.6 **a** The refractive index and absorption coefficient of blood; **b** the rate of blood flow in the capillary vessel at an instant t in the vicinity of the point $x' = 0.0001$, $y' = 0.0001$ (**b**)

In Fig. 4.5a and b, the dependence of the intensity of radiation on the coordinate system is shown for a multilayer absorbing and scattering medium, which models human skin, for different values of the refractive index of blood. It should be noted that, in comparison with the values of the refractive index of blood given in [9], our model is rather stable and is sensitive to a change in this parameter up to the fifth decimal place. This makes possible a more exact diagnostics of various pathological processes related to changes in the electrophysical properties of blood.

The dependences of the intensity of the laser radiation on the refractive index and absorption coefficient for the system of blood vessels in the upper dermis are presented in Fig. 4.6a.

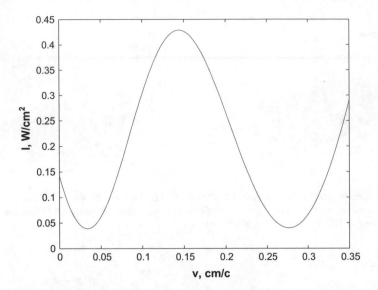

Fig. 4.7 The rate of blood flow in the capillary vessel at an instant t(thickness of the epidermis is 65 μm; thickness of the upper dermis is 600 μm; thickness of the blood is 85 μm, refraction coefficients epidermis, upper dermis, blood and lower dermis are $1.4500 + j \cdot 10^{-5}$; $1.400 + j \cdot 10^{-5}$; $1.300 + j \cdot 10^{-5}$; $1.400 + j \cdot 10^{-5}$)

In Fig. 4.6b, the intensity of the laser radiation is shown as a function of the rate of blood flow in a capillary at the time instant t in the vicinity of some point x', y'. These quantitative estimates allow one to determine a change in the rate of the blood flow in the capillary bed, which makes possible the study of physiological processes occurring in skin.

Note that the principle of measuring blood flow velocity is based on the interference of a stationary laser beam of light on a system of rough layers, one of which (blood-bearing) can be represented as moving with the blood flow velocity v_x.

Therefore, the movement of the blood will lead to a pseudoperiodic time variation of the intensity with a characteristic time on the order of

$$\frac{\delta x}{v_x},$$

where δx is the characteristic horizontal size of roughness. Figures 4.7 and 4.8 illustrate the rate of blood flow in the capillary vessel at an instant t.

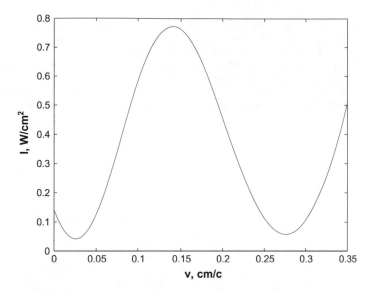

Fig. 4.8 The rate of blood flow in the capillary vessel at an instant t(thickness of the epidermis is 65 μm; thickness of the upper dermis is 600 μm; thickness of the blood is 70 μm, refraction coefficients epidermis, upper dermis, blood and lower dermis are $1.4500 + j \cdot 10^{-5}$; $1.400 + j \cdot 10^{-5}$; $1.300 + j \cdot 10^{-5}$; $1.400 + j \cdot 10^{-5}$)

The dependences presented can be used for the prediction of changes in the optical properties of blood and in the rate of the blood flow in the capillary bed caused by various biophysical, biochemical, and physiological processes. Similar dependences can be calculated for lasers with other parameters. The quantitative estimates obtained can be applied to processing and interpreting of experimental data.

References

1. D.Y. Paithankar, V.E. Ross, B.A. Saleh, M.A. Blair, B.S. Graham, Acne treatment with a 1450 nm wavelength laser and cryogen spray cooling. Lasers Surg. Med. **31**(2), 106–114 (2002)
2. A.Yu. Seteykin, The model for calculating the temperature fields generated by the laser radiation on multilayer biological tissue. J. Opt. Technol. **72**(7), 42–47 (2005)
3. C.T.W. Lahaye, M.J.C. van Gemert, Optimal laser parameters for port wine stain therapy: a theoretical approach. Phys. Med. Biol. **30**(6), 573–588 (1985)
4. L.G. Astafeva, G.I. Zheltov, Modelling of of the heating process of blood vessels by laser radiation. Opt. Spectrosc. **90**(2), 287–292 (2001)
5. L.E. Dolotov, YuP Sinichkin, V.V. Tuchin, S.R. Utz, G.B. Altshuler, I.V. Yaroslavsky, Design and evaluation of a novel portable Erythema-Melanin-Meter. Lasers Surg. Med. **34**, 127–135 (2004)
6. I.V. Meglinski, Simulation of the reflectance spectra of optical radiation from a randomly inhomogeneous multilayer strongly scattering and absorbing light environments using the Monte Carlo. Quantum Electron. **31**(12), 1101–1107 (2001)

7. I.V. Fedosov, V.V. Tuchin, in the spatial-temporal correlation of the intensity of the speckle field formed by the scattering of coherent radiation focused on the capillary flow of fluid containing scattering particles. Opt. Spectrosc. **93**(3), 473–477 (2002)
8. K.G. Kulikov, A.I. Radin, An electrodynamic model of the optical characteristics of blood and capillary blood flow rate. Opt. Spectrosc. **96**(4), 613–625 (2004)
9. A.N. Korolevich, A.J. Khairullina, L.P. Shubochkin, The scattering matrix for monolayer of "soft" particles when there is dense packing. Opt. Spectrosc. **68**(2), 403–409 (1990)

Chapter 5
Study of the Optical Characteristics of a Biotissue with Large-Scale Inhomogeneities

Abstract We construct the electrodynamic model which makes it possible to vary the electrophysical parameters of a biological structure in calculations with allowance for roughness (real and imaginary parts of the refractive index of the epidermis, the upper layer of the derma, and blood) and to establish the dependences between these parameters and the biological properties of blood under the action of laser radiation in vivo.

5.1 Introduction

It should be noted that the study of propagation of light in randomly inhomogeneous media is mainly based on the classical methods of the transport theory. However, the application of the radiation transport theory is not always effective in the study of propagation of light in randomly inhomogeneous media (in particular, biological media). It is well known that most biological surfaces are not planar but are rather loose randomly rough media, in which the size of roughness are larger than the wavelength of radiation illuminating them. The roughness of the surfaces affects the characteristics of propagation and scattering of waves. A wave incident on a rough surface is not only reflected specularly, but is also scattered in all other directions. The spatial000 parameters of the light beam interacting with a rough interface in this case obviously change to a certain extent as compared to the case when radiation is incident on a smooth surface. However, the classical transport theory fails to indicate how the spatial parameters of the beam may change when it intersects the rough interface between two media [1]. Thus, the disregard of rough boundaries in the transport theory requires the application of classical methods of the theory of diffraction of electromagnetic waves from randomly rough surfaces. In this connection, it is important to. investigate the optical characteristics of the biological structure taking roughness (when the characteristic size of roughness on the surface is much larger than the wavelength) with the help of the classical methods of diffraction theory. A number of publications in which light scattering from a rough surface was studied [1–4] are worth men tioning. For example, the scattering of light from a rough surface with random Gaussian fluctuations of roughness was studied

© Springer International Publishing AG, part of Springer Nature 2018
K. Kulikov and T. Koshlan, *Laser Interaction with Heterogeneous Biological Tissue*, Biological and Medical Physics, Biomedical Engineering,
https://doi.org/10.1007/978-3-319-94114-1_5

in [1]. The case of coarse roughness was considered, when the parameters (standard deviation and correlation radius) are much larger than the wavelength. In [3], light scattering from an anisotropic rough surface was also considered. Scattering of light from a rough cylindrical surface was studied in [4]. In [2], scattering of light from a rough dielectric surface was analyzed (both theoretically and numerically).

Here, we construct an electrodynamic model which makes it possible to vary the electrophysical parameters of a biological structure in calculations with allowance for roughness (real and imaginary parts of the refractive index of the epidermis, the upper layer of the derma, and blood) and to establish the dependences between these parameters and the biological properties of blood under the action of laser radiation for case in vivo. The problem consists of three consecutive stages. At the first stage, the problem of light scattering from a rough boundary is solved and the coefficient of reflection of a plane wave from a smoothly irregular layer simulating the given biological medium is determined taking into account the roughness of the interface in the case when the size of the roughness is larger than the wavelength of radiation illuminating them.

At the second stage, we solve the problem of reflection of a Gaussian beam with an arbitrary cross section. The problem is solved by expanding the fields of counter propagating waves in plane waves in the domain of medium 1 and their reflection by layer 2 and inverse transformation followed by the Huygens-Fresnel integral transformation to obtain the field in the initial reference cross section (see Chap. 4). At the third stage, the dependence of the radiation intensity on the refractive index is determined for a system of blood vessels in the upper layer of the dermis and the effect of roughness on the electrophysical characteristics of the biological sample being simulated is analyzed. The structure simulated consists of three regions with different refractive indices (epidermis, upper layer of the dermis, and blood vessel) illuminated by a laser beam for case in vivo.

The chapter is based on the results of the [5, 6].

5.2   Scattering of a Plane Wave from the Rough Surface

The surfaces of real bodies (in particular, in biology) are not perfectly smooth to a certain extent. For this reason, reflection and refraction of the waves at these surfaces are accompanied by the effects which are not observed for smooth surfaces. Rigorous methods for solving the problem in the case of a rough surface do not exist. The problem can be solved only approximately under certain constraints imposed on the size and shape of roughness. For calculating the scattered field, the small-perturbation method and the Kirchhoff (tangential plane) method are used. In this work, we will use the Kirchhoff method for calculating the scattered field. To solve the formulated problem, we will use the Kirchhoff approximation. To apply the Kirchhoff method correctly, we make the assumption concerning the smoothness of inhomogeneities. At each point, the wave field can be represented as the sum of incident field (E_{inc}) and reflected field (E_{ref}).

Let us write the expressions for the field scattered by a certain smooth rough surface $z = H(x, y)$ in the Kirchhoff approximation. We select a certain region S of this surface, whose linear size is much larger than the mean size of the roughness, which in turn is much larger than the wavelength. We assume that there are no elements of the surface shadowed from the incident wave or scattered wave.

Let us suppose that a plane s- or p- monochromatic wave is incident on a rough surface; the unit wave amplitude has the form

$$E_{inc}(r) = e^{-i\mathbf{k_1 r}}.$$

The observation will be carried out in the Fraunhofer zone of domain S and in the direction of wave vector $\mathbf{k}$. In this zone, the elementary waves of all elements of the scattering domain can be treated as plane waves.

Fields $\mathbf{E}$ and $\mathbf{H}$ can be expressed in terms of certain scalar function (e.g., E) that satisfies the equation

$$\Delta E + k^2 E = 0 \tag{5.1}$$

with boundary conditions of the form [7, 8]

$$E_{ref}|_{z=H(x,y)} = (1 + V)E_{inc}|_{z=H(x,y)}, \tag{5.2}$$

$$\frac{\partial E_{ref}}{\partial \mathbf{n}}|_{z=H(x,y)} = (1 - V)\frac{\partial E_{inc}}{\partial n}|_{z=H(x,y)}, \tag{5.3}$$

where $k^2 = \omega^2 \varepsilon_0 \mu_0$, V is the reflection coefficient depending on physical parameters of the medium, and $\mathbf{n}$ is the unit vector of the outward normal. It should be borne in mind that the formulas for the reflection coefficient for the s- or p- polarization are different.

It should be noted that using the Kirchhoff method, we solve not the boundary-value problem of diffraction, but a simpler problem that basically differs from it (i.e., the problem with a preset discontinuity of the field and of its normal derivative on the surface). Thus, in contrast to the perturbation method considered in Chap. 9, in which the applicability limits of the results can be indicated for a wide class of special cases and the next terms of the expansion can be calculated from the known small parameters, the results obtained using the Kirchhoff method cannot be treated as the expansion of the exact solution into a series in a small parameter (e.g., the ratio of the wavelength to the characteristic linear size of the body at which diffraction takes place).

It is well known that the values of E inside the domain are connected with E and $\frac{\partial E}{\partial n}$ on the surface bounding this domain by the Green formula

$$E(\mathbf{r}) = \oint_S \left[E(\mathbf{r'})\frac{\partial G(\mathbf{r}, \mathbf{r'})}{\partial \mathbf{n}} - \frac{\partial E(\mathbf{r'})}{\partial \mathbf{n}}G(\mathbf{r}, \mathbf{r'}) \right] dS, \tag{5.4}$$

where $G(\mathbf{r}, \mathbf{r'})$ is the Green function, which has the form

$$G(\mathbf{r}, \mathbf{r}') = \frac{1}{4\pi} \frac{e^{-ikR}}{R}, \quad R = |\mathbf{r} - \mathbf{r}'|.$$

Taking into account expression (5.4), we obtain

$$E(\mathbf{r}) = E_{inc}(\mathbf{r}) + \oint_S \left[E_{ref}(\mathbf{r}') \frac{\partial G(\mathbf{r}, \mathbf{r}')}{\partial \mathbf{n}} - \frac{\partial E_{ref}(\mathbf{r}')}{\partial \mathbf{n}} G(\mathbf{r}, \mathbf{r}') \right] dS \qquad (5.5)$$

$$\oint_S \left[E_{inc}(\mathbf{r}') \frac{\partial G(\mathbf{r}, \mathbf{r}')}{\partial \mathbf{n}} - \frac{\partial E_{inc}(\mathbf{r}')}{\partial \mathbf{n}} G(\mathbf{r}, \mathbf{r}') \right] dS = 0. \qquad (5.6)$$

Expression (5.6) implies that all sources of the field lie within the surface. Subtracting expression (5.6) from (5.5), we arrive to the formula

$$E(\mathbf{r}) = E_{inc}(\mathbf{r}) + \oint_S \left[(E_{inc}(\mathbf{r}') - E_{ref}(\mathbf{r}')) \frac{\partial G(\mathbf{r}, \mathbf{r}')}{\partial \mathbf{n}} \right] -$$

$$- \left[\frac{\partial E_{inc}(\mathbf{r}')}{\partial \mathbf{n}} - \frac{\partial E_{ref}(\mathbf{r}')}{\partial \mathbf{n}} \right] G(\mathbf{r}, \mathbf{r}') dS. \qquad (5.7)$$

Substituting the value of the field and its derivative into this expression, we obtain

$$E(\mathbf{r}) = E_{inc}(\mathbf{r}) + \oint_S V \left[E_{inc}(\mathbf{r}') \frac{\partial G(\mathbf{r}, \mathbf{r}')}{\partial \mathbf{n}} - \frac{\partial E_{inc}(\mathbf{r}')}{\partial \mathbf{n}} G(\mathbf{r}, \mathbf{r}') \right] dS. \qquad (5.8)$$

We will henceforth consider only the scattered field defined as

$$E_{scat}(\mathbf{r}) = \oint_S V \left[E_{inc}(\mathbf{r}') \frac{\partial G(\mathbf{r}, \mathbf{r}')}{\partial \mathbf{n}} - \frac{\partial E_{inc}(\mathbf{r}')}{\partial \mathbf{n}} G(\mathbf{r}, \mathbf{r}') \right] dS. \qquad (5.9)$$

Substituting into this formula the approximate expression for the derivative of e^{ikR}/R with respect to n and the corresponding approximate expression for e^{ikR}/R, which are defined as [9]

$$\frac{\partial G}{\partial \mathbf{n}} \cong \frac{1}{4\pi} \frac{e^{-ikr}}{r} \frac{\partial e^{i(\mathbf{k}, \mathbf{r})}}{\partial \mathbf{n}} = \frac{1}{4\pi} \frac{e^{-ikr}}{r} i(\mathbf{n}, \mathbf{k}) e^{i(\mathbf{k}, \mathbf{r})},$$

$$\frac{\partial E_{inc}}{\partial \mathbf{n}} \cong \frac{1}{4\pi} \frac{e^{-ikr}}{r} \frac{\partial e^{-i(\mathbf{k_1}, \mathbf{r})}}{\partial \mathbf{n}} =$$

$$= \frac{1}{4\pi} \frac{e^{-ikr}}{r} i(\mathbf{n}, \mathbf{k_1}) e^{-i(\mathbf{k_1}, \mathbf{r})}, \quad \frac{e^{ikR}}{R} \cong \frac{1}{4\pi} \frac{e^{ikr}}{r} e^{i(\mathbf{k}, \mathbf{r})}, \qquad (5.10)$$

we obtain

$$E_{scat}(r) = i \frac{e^{-ikr}}{4\pi r} \mathbf{q} \oint_S V \left[\mathbf{n} e^{i(\mathbf{q}, \mathbf{r})} \right] dS, \qquad (5.11)$$

where

$$\mathbf{q} = \mathbf{k} - \mathbf{k}_1, \quad dS = \frac{dxdy}{\alpha}, \quad \mathbf{n} = \left(\alpha \frac{\partial H}{\partial x}, \alpha \frac{\partial H}{\partial y}, -\alpha \right), \quad (5.12)$$

$$\alpha = \frac{1}{\sqrt{(1 + (\frac{\partial H}{\partial x})^2 + (\frac{\partial H}{\partial y})^2}}.$$

This gives

$$\mathbf{n}dS = \left[\frac{\partial H}{\partial x}, \frac{\partial H}{\partial y}, -1 \right] dxdy, \quad (\mathbf{q}, \mathbf{r}) = q_x x + q_y y + q_z H(x, y), \quad (5.13)$$

$$\frac{\partial e^{i(\mathbf{q},\mathbf{r})}}{\partial x} = i \left[q_x + q_z \frac{\partial H}{\partial x} \right] e^{i(\mathbf{q},\mathbf{r})}, \quad (5.14)$$

$$\frac{\partial e^{i(\mathbf{q},\mathbf{r})}}{\partial y} = i \left[q_y + q_z \frac{\partial H}{\partial y} \right] e^{i(\mathbf{q},\mathbf{r})}, \quad (5.15)$$

$$\mathbf{n}e^{i(\mathbf{q},\mathbf{r})} dS = \left[\frac{\partial e^{i(\mathbf{q},\mathbf{r})}}{i\partial x} - q_x e^{i(\mathbf{q},\mathbf{r})}, \frac{\partial e^{i(\mathbf{q},\mathbf{r})}}{i\partial y} - q_y e^{i(\mathbf{q},\mathbf{r})}, -q_z e^{i(\mathbf{q},\mathbf{r})} \right] \frac{dxdy}{q_z}. \quad (5.16)$$

Substituting expression (5.16) into (5.11) and taking into account relations (5.12)–(5.15), we obtain

$$E_{scat}(r) = i \frac{e^{-ikr}}{4\pi r} \oint_{S_0} V \left[\frac{-q^2}{q_z} + \frac{1}{iq_z} q_x \frac{\partial}{\partial x} + \frac{1}{iq_z} q_y \frac{\partial}{\partial y} \right] e^{i(\mathbf{q},\mathbf{r})} dxdy, \quad (5.17)$$

where $q_x = -k(\sin \theta_s \cos \varphi_s - \sin \theta_i)$, $q_y = -k \sin \theta_s \sin \varphi_s$, $q_z = -k(\cos \theta_i + \cos \theta_s)$, θ_s is the scattering angle, θ_i is the angle of incidence and φ_s is the azimuthal angle.

Note that in formula (5.17) we have passed from integration over surface S to integration over its projection S_0 onto the plane $z = 0$. Let us write expression (5.17) in the form

$$E_{scat}(x, y) = -i \frac{e^{-ikr}}{4\pi r} \frac{q^2}{q_z} \left[\oint_{S_0} V \left[e^{i(q_x x + q_y y + q_z H(x,y))} \right] dxdy \right] +$$

$$+ i \frac{e^{-ikr}}{4\pi r} \left[\oint_{S_0} V \left[\frac{q_x}{q_z} \left[q_x + q_z \frac{\partial H(x, y)}{\partial x} \right] + \frac{q_y}{q_z} \left[q_y + \frac{\partial H(x, y)}{\partial y} \right] \right] \right] \times \quad (5.18)$$

$$\times e^{i(q_x x + q_y y + q_z H(x,y))} dxdy.$$

It should be noted that the second part of formula (5.18) gives the edge effect.

5.3   The Scattered Field on the Fractal Surface

It should be noted that many biological tissues (in particular, dermis) exhibit optical inhomogeneity [10, 11]; in this case, the surface of the outer dermis of the biological structure being modeled can be described by the following 2D range-limited Weierstrass function:

$$H_2(x, y) = \sigma \sqrt{\left[\frac{2(1 - q_2^{2(D_2-3)})}{M_2(1 - q_2^{2(D_2-3)N})} \right]} \times \tag{5.19}$$

$$\times \sum_{n=0}^{N-1} q_2^{(D_2-3)n} \sum_{m=1}^{M} c_2 \sin \left[K_2 q_2^n \left[a_2 x \cos \frac{2\pi m}{M_2} + b_2 y \sin \frac{2\pi m}{M_2} \right] + \varphi_{nm} \right],$$

where a_2, b_2, c_2 are arbitrary constants obeying the conditions $a_2 \ll 1$, $b_2 \ll 1$, $c_2 \ll 1$.

In formula (5.19) $q_2 > 1$ is the parameter of the spatial-frequency scaling, D_2 is the fractal dimension, K_2 is the principal spatial wave number, N_2 and M_2 are the numbers of harmonics, φ_{nm} is the arbitrary phase which is distributed uniformly in the interval $[-\pi, \pi]$, and σ is the standard deviation. Function $H_2(x, y)$ is self-similar and has derivatives. The surface based on this function has many scales, and the roughness can change depending on the scale under consideration.

To describe the rough surface numerically, parameters like the correlation interval, the standard deviation, and spatial autocorrelation coefficient are normally used. The possibility of using these statistical parameters for estimating the effect of the fractal dimension and other parameters on the roughness of the surface was considered in [12]. For this purpose, the dependence of the mean correlation interval on D for various values of q and the dependence of the mean correlation interval on q for various values of D are investigated numerically. It is shown that inhomogeneities of the fractal surface are mainly controlled by quantity D. Note that the fractal surface presumes the presence of roughness of all scales relative to the wavelength of the scattered wave. The features of scattering of waves by the fractal surface are determined by the fact the surface is not differentiable; thus, the fractal front, which is not differentiable, has no normal. However, the chords connecting the values of the characteristic heights of roughness at certain distances have a finite root-mean-square slope. In this case, the hypothesis of a fractal chaotic surface is introduced; it is equal to the length over which the slopes of the surface are close to unity [13]. Thus, two scattering models have been adopted at present; the first is the model with fractal heights, while the second is the model with fractal slopes of roughness. In the second model, we note that it is once differentiable and has a slope that varies continuously from point to point. This allows us to analyze our model in the geometrical optics approximation.

Substituting expression (5.19) into (5.18), we obtain

$$
E_{scat}(x, y) = -i \frac{e^{-ikr}}{4\pi r} \frac{q^2}{q_z} \oint_{S_0} V \exp\left[q_x x + q_y y + q_z c \sum_{n=0}^{N-1} q_2^{(D_2-3)n} \right] \times
$$

$$
\times \exp\left[\sum_{m=1}^{M} c_2 \sin K_2 q_2^n \left[a_2 x \cos \frac{2\pi m}{M_2} + b_2 y \sin \frac{2\pi m}{M_2} \right] + \varphi_{nm} \right] dx dy =
$$

$$
= -i \frac{e^{-ikr}}{4\pi r} \frac{q^2}{q_z} \oint_{S_0} V \prod_{n=0}^{N-1} \prod_{m=1}^{M} \exp\left[q_x x + q_y y + q_z c q_2^{(D_2-3)n} \right] \times
$$

$$
\times \exp\left[c_2 \sin K_2 q_2^n \left[a_2 x \cos \frac{2\pi m}{M_2} + b_2 y \sin \frac{2\pi m}{M_2} \right] + \varphi_{nm} \right] dx dy, \qquad (5.20)
$$

where

$$
c = \sigma \sqrt{\left[\frac{2(1 - q_2^{2(D_2-3)})}{M_2(1 - q_2^{2(D_2-3)N})} \right]}.
$$

We will use the representation of the uth-order Bessel function of the first kind in the form of a power series

$$
e^{ikz \sin \phi} = \sum_{u=-\infty}^{\infty} J_u(z) e^{iu\phi}. \qquad (5.21)
$$

We substitute expression (5.21) into (5.20), which gives

$$
E_{scat}(x, y) = -i \frac{e^{-ikr}}{4\pi r} \frac{q^2}{q_z} \oint_{S_0} V \prod_{n=0}^{N-1} \prod_{m=1}^{M} \sum_{u=-\infty}^{\infty} J_{u_{nm}}(q_z c q_2^{(D_2-3)n}) \times \qquad (5.22)
$$

$$
\times \exp\left[q_x x + q_y y + iu \left[c_2 K_2 q_2^n \left[a_2 x \cos \left[\frac{2\pi m}{M_2} \right] + b_2 y \sin \left[\frac{2\pi m}{M_2} \right] \right] + \varphi_{nm} \right] \right].
$$

We can write this expression in the form

$$
E_{scat}(x, y) = -i \frac{e^{-ikr}}{4\pi r} \frac{q^2}{q_z} \oint_{S_0} V \sum_{u_{M,N-1}=-\infty}^{\infty} \prod_{n=0}^{N-1} \prod_{m=1}^{M} J_{u_{nm}}(q_z c q_2^{(D_2-3)n}) \times
$$

$$
\times \exp\left[c_2 i K_2 \left[\sum_{n=0}^{N-1} q_2^n \sum_{m=1}^{M} u_{nm} a_2 x \cos \left[\frac{2\pi m}{M_2} \right] \right] + q_x x \right] \times \qquad (5.23)
$$

$$\times \exp\left[c_2 i K_2 \left[\sum_{n=0}^{N-1} q^n \sum_{m=1}^{M} u_{nm} b_2 y \sin\left[\frac{2\pi m}{M_2}\right]\right] + q_y y\right] \exp\left[i \sum_{n=0}^{N-1}\sum_{m=1}^{M} u_{nm}\varphi_{nm}\right].$$

If we consider scattering from a finite area element of size $2L_x \times 2L_y$ for $-L_x \le x \le L_x$ and $-L_y \le y \le L_y$, then taking expression (5.23) into account analogously to [13], we obtain the following expression for the scattered field:

$$E_{scat}(x, y) = -i \frac{e^{-ikr}}{\pi r}\frac{q^2}{q_z} \sum_{u_{M,N-1}=-\infty}^{\infty} \prod_{n=0}^{N-1}\prod_{m=1}^{M} J_{u_{nm}}(q_z c q_2^{(D_2-3)n}) \times$$

$$\times \exp\left[i \sum_{n=0}^{N-1}\sum_{m=1}^{M} u_{nm}\varphi_{nm}\right] \frac{\sin(L_x \vartheta_x)}{\vartheta_x}\frac{\sin(L_y \vartheta_y)}{\vartheta_y} + \Psi_k, \qquad (5.24)$$

where Ψ_k gives the edge effect, and,

$$\vartheta_x = q_x + c_2 K_2 \left[\sum_{n=0}^{N-1} q_2^n \sum_{m=1}^{M} u_{nm} a_2 x \cos\left[\frac{2\pi m}{M_2}\right]\right],$$

$$\vartheta_y = q_y + c_2 K_2 \left[\sum_{n=0}^{N-1} q_2^n \sum_{m=1}^{M} u_{nm} b_2 y \sin\left[\frac{2\pi m}{M_2}\right]\right].$$

5.4   Reflection of a Plane Wave from a Layer with Allowance for Surface Roughness

Having derived the expression for the field scattered by a certain smooth uneven surface $z = H(x, y)$ in the Kirchhoff approximation in the case when the characteristic size of roughness on the surface considerably exceeds the wavelength, we consider the problem of reflection of a plane wave from a layer with a slowly varying thickness taking the roughness into account.

Let us consider the following optical scheme. The system consists of three domains with difference refractive indices(epidermis, the upper layer of the dermis, blood vessel). To attain the best agreement between the structure and the actual object under investigation, we represent the interfaces between the layers of the model medium in the form of certain surfaces $z_i = H_i(x, y)$, $i = \overline{1, 2}$, where $z_1 = H_1(x, y)$ is defined by expression (4.1) and $z_2 = H_2(x, y)$ is defined by expression (5.19).

Let us suppose that a plane s- or p- polarized wave is incident on the layer at an angle θ. We consider only the case of the p polarization. We must find the reflected field.

We will seek the reflected field in the form of waves with slowly varying amplitudes and rapidly oscillating phases

$$E_1 = \exp\left(\frac{i}{\varepsilon}\tau_{inc}(\xi_1, \xi_2, \xi_3)\right) + \exp\left(\frac{i}{\varepsilon}\tau_{1ref}(\xi_1, \xi_2, \xi_3)\right) A(\xi_1, \xi_2, \xi_3), \quad (5.25)$$

$$E_2 = \exp\left(\frac{i}{\varepsilon}\tau_{2tr}(\xi_1, \xi_2, \xi_3)\right) B^+(\xi_1, \xi_2, \xi_3) +$$

$$+ \exp\left(\frac{i}{\varepsilon}\tau_{3ref}(\xi_1, \xi_2, \xi_3)\right) B^-(\xi_1, \xi_2, \xi_3), \quad (5.26)$$

$$E_3 = \exp\left(\frac{i}{\varepsilon}\tau_{3elap}(\xi_1, \xi_2, \xi_3)\right) C^+(\xi_1, \xi_2, \xi_3) +$$

$$+ \exp\left(\frac{i}{\varepsilon}\tau_{4ref}(\xi_1, \xi_2, \xi_3)\right) C^-(\xi_1, \xi_2, \xi_3) + E_{scat}(\xi_1, \xi_2), \quad (5.27)$$

where $E_{scat}(\xi_1, \xi_2)$ in the general form is defined by expression (5.24).

$$E_4 = \exp\left(\frac{i}{\varepsilon}\tau_{4elap}(\xi_1, \xi_2, \xi_3)\right) D(\xi_1, \xi_2, \xi_3), \quad (5.28)$$

and $\tau_{1inc}, \tau_{1ref}, \tau_{2elap}, \tau_{3ref}, \tau_{3elap}, \tau_{4ref}, \tau_{4elap}$ are defined in Chap. 4.

Amplitudes A, $B^{\pm}$, $C^{\pm}$ and D are sought in the form of power series on for small parameter ε_x, ε_y, the expressions for the amplitudes are defined analogously to the method described in Chap. 4.

Substitution of expressions (5.25)–(5.28) into (4.6)–(4.11) generates a recurrence system of equations. For the reflected field, this system leads to reflection coefficient A taking into account the roughness at the interface with the medium being simulated in the case when the characteristic size of roughness on the surface is much larger then the wavelength of incident radiation.

$$1 + A_{\widehat{00}} = B_{\widehat{00}}^{+} + B_{\widehat{00}}^{-},$$

$$B_{\widehat{00}}^{+} \exp(-h_1 i k'_{2z}) + B_{\widehat{00}}^{-} \exp(h_1 i k'_{3z}) = C_{\widehat{00}}^{+} \exp(-h_1 i k'_{3z}) + C_{\widehat{00}}^{-} \exp(h_1 i k'_{4z})$$

$$+ E_{scat}(\xi_1, \xi_2, \xi_3)|_{\xi_3 = \varepsilon h_1(\xi_1, \xi_2)},$$

$$C_{\widehat{00}}^{+} \exp(-h_2 i k'_{3z}) + C_{\widehat{00}}^{-} \exp(h_2 i k'_{4z}) + E_{scat}(\xi_1, \xi_2, \xi_3)|_{\xi_3 = \varepsilon h_1(\xi_1, \xi_2)} =$$

$$D_{\widehat{00}}^{+} \exp(-h_2 i k'_{4z}) + D_{\widehat{00}}^{-} \exp(h_2 i k'_{5z}),$$

$$D_{\widehat{00}+} \exp(-h_3 i k'_{4z}) + D_{\widehat{00}}^{-} \exp(h_3 i k'_{5z}) = E_{\widehat{00}} \exp(-h_3 i k_{5z})$$

$$\frac{1}{n_1^2}\left(-ik_{1z}+ik_{2z}\widehat{A_{00}}\right)=\frac{1}{n_2^2}\left(-ik_{2z}'\widehat{B_{00}}^{+}+ik_{3z}'\widehat{B_{00}}^{-}\right),$$

$$\frac{1}{n_2^2}\left(-ik_{2z}'\widehat{B_{00}}^{+}\exp(-h_1ik_{2z}')+ik_{3z}'\widehat{B_{00}}^{-}\exp(h_1ik_{3z}')\right)=$$

$$=\frac{1}{n_3^2}\left(-ik_{3z}'\widehat{C_{00}}^{+}\exp(-h_1ik_{3z}')+ik_{4z}'\widehat{C_{00}}^{-}\exp(h_1ik_{4z}')\right)+$$

$$+\frac{1}{n_3^2}\frac{\partial}{\partial\xi_3}E_{scat}(\xi_1,\xi_2,\xi_3)|_{\xi_3=\varepsilon h_1(\xi_1,\xi_2)},$$

$$\frac{1}{n_3^2}\left(-ik_{3z}'\widehat{C_{00}}^{+}\exp(-h_2ik_{3z}')+ik_{4z}'\widehat{C_{00}}^{-}\exp(h_2ik_{4z}')\right)+$$

$$+\frac{1}{n_3^2}\frac{\partial}{\partial\xi_3}E_{scat}(\xi_1,\xi_2,\xi_3)|_{\xi_3=\varepsilon h_1(\xi_1,\xi_2)}=$$

$$=\frac{1}{n_4^2}\left(-ik_{4z}'\widehat{D_{00}}^{+}\exp(-h_2ik_{4z}')+ik_{5z}'\widehat{D_{00}}^{-}\exp(h_2ik_{5z}')\right),$$

$$\frac{1}{n_4^2}\left(-ik_{4z}'\widehat{D_{00}}^{+}\exp(-h_2ik_{4z}')+ik_{5z}'\widehat{D_{00}}^{-}\exp(h_2ik_{5z}')\right)=$$

$$=\frac{1}{n_5^2}\left(-ik_{5z}\widehat{E_{00}}\exp(-h_3ik_{5z})\right).$$

The expression for the reflection of a Gaussian beam with an arbitrary cross section is defined by expression (4.67). The expressions for the reflected H field are derived in a similar manner.

The radiation intensity is defined by expression (4.69) and Substitution of expressions (4.67) into (4.69) and thus, at the given stage, we have derived the expressions that make it possible to determine the explicit dependence of the intensity of laser radiation on the refractive index and the absorption coefficient for a system of blood vessels in the upper layer of the dermis. Further analysis of the above dependences will be carried out using numerical methods.

5.5 Numerical Calculations for a Model Medium and Conclusions

Let us consider the optical system being simulated. The system consists of three domains with different refractive indices (epidermis, the outer layer of the dermis, and a blood vessel). The system has the following parameters. Refractive indices of the layers are $n_2^\circ = 1.33$, $n_3^\circ = 1.35$, $n_4^\circ = 1.35$ and the characteristic thicknesses of the layers are $d_1 = 65 \cdot 10^{-6}$, $d_2 = 565 \cdot 10^{-6}$, $n_1^\circ = 1$, $\chi_1 = 0$, $\chi_2 = \chi_3 = \chi_4 = 10^{-5}$, $a_1 = -0.0987$, $b_1 = 0.09920$, $c_1 = 0.07749$, $a_2 = 0.007$, $b_2 = 0.0089$, $c_2 = 0.0089$, wavelength is $\lambda = 0.63\,\mu\text{m}$ (center of the line of a He$-$Ne laser), $q_2 = 1.01$, $K_2 = 6$, $N_2 = 2$, $M_2 = 3$, $D_2 = 2.9$, $\sigma = 1$; i.e., function $H_2(x, y)$ is normalized to $\sigma = 1$. Note that the values of parameters a_i, b_i and c_i, $i = \overline{1, 2}$ are selected so that the shape of the surface corresponds best to the shape of the interface with the corresponding layer in the structure of the normal human dermis. In numerical calculations, the edge effect was not taken into account.

Figure 5.1 shows the dependence of the intensity of scattered field on fractal dimension D. The curve in Fig. 5.1 shows that the scattering intensity increases, then the relief of the surface becomes more complicated (fractal dimension D increases). This effect can be explained by an increasing contribution of secondary scattering from fine roughness as compared to that in the case of a smoother surface.

Figure 5.2 shows the radiation intensity for specific values of electrophysical parameters and q_2, K_2, D_2, N_2, M_2 a_i, b_i, c_i, $i = \overline{1, 2}$ for a multilayer absorbing medium and scattering medium simulating the human dermis. It can be seen from Fig. 5.2 that the Gaussian beam splits.

Figures 5.3a, b show the dependences of the laser radiation intensity on the refractive index and absorption coefficient for a system of blood vessels in the outer layer of the dermis for various absorption coefficients of the epidermis and dermis.

Fig. 5.1 Dependence of the squared modulus of the scattered field on fractal dimension D

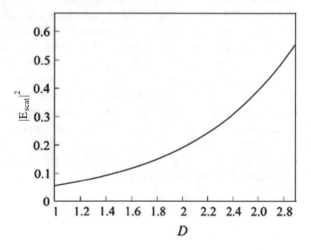

Fig. 5.2 Radiation intensity
for specific values of
parameters and for $\theta = 0^0$,
$\varphi = 0^0$, $\psi = 0^0$

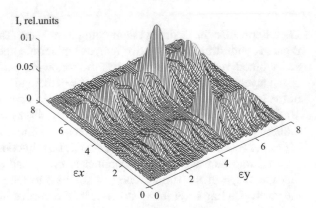

It should be noted that the model constructed here is quite sensitive to changes in
the electrophysical parameters of the biological structure being simulated (in particular, to the absorption coefficient). The model constructed here makes it possible to
vary the inhomogeneities of a rough surface, the electrophysical parameters of the
biological samples under investigation, and the geometrical parameters and to establish the dependences between these parameters and the biological properties of the
biotissue being simulated. Thus, this model can be used for measuring the spectral
differences between the normal and pathological tissues in vivo experiments taking
into account large-scale inhomogeneities, aimed at the development of a spectral
autograph for determining pathological changes in the biological samples, which are
associated with a variation of the electrophysical properties of the epidermis, outer
dermis, and blood.

Analogous dependences can be obtained for lasers with other parameters and
can be used for processing of the experimental absorption curves of the biological
structures under investigation with regard to large-scale inhomogeneities.

The dependences given above can be used for predicting changes in the optical
properties of blood in the capillary channel, which are associated with various biophysical, biochemical, and physiological processes in the blood; these dependences
can also be calculated for lasers with other parameters; and the quantitative estimates
can also be applicable for experimental data processing and interpretation.

The model considered here will be extended by considering scattering from large-
scale inhomogeneities with allowance for multiple scattering to obtain a more adequate description of reflection from the rough interface in simulated biological
structures of various morphologies.

Fig. 5.3 Dependence of the laser radiation intensity on refractive index n_4 and absorption coefficient χ_4 for a system of blood vessels in the outer layer of the dermis for the absorption coefficient of the epidermis and dermis $\chi_4 = 10^{-4}$; **b** the same as in **a** but for $\chi_4 = 10^{-5}$

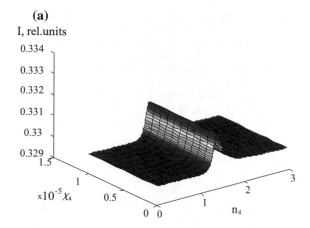

(a)

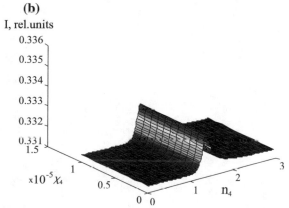

(b)

References

1. D.A. Rogatkin, The scattering of electromagnetic waves by randomly rough surfaces as a boundary value problem of interaction of laser radiation with light-scattering materials and media. Opt. Spectrosc. **97**(3), 484–493 (2004)
2. J.A. Sanchez-Gil, M. Nieto-Vesperinas, Light scattering from random rough dielectric surfaces. J. Opt. Soc. Am. A. **8**(8), 1270–1286 (1991)
3. Li Hai-Xia, Cheng Chuan-Fu, Light scattering of rough orthogonal anisotropic surfaces with secondary most probable slope distributionschin. Phys. Lett. **28**, N8 (2011)
4. Y.Y. Fan, V.M. Huynh, Light scattering from periodic rough cylindrical surfaces. Appl. Optics. **32**(19), 3452–3458 (1993)
5. K.G. Kulikov, Simulation of the electrophysical characteristics of a biological tissue with allowance for large–scale inhomogeneities. Tech. Phys. **57**(7), 907–914 (2012)
6. K.G. Kulikov, Mathematical modeling of optical properties of biological structures, taking into account large-scale inhomogeneities, in *SPIE Photonics Europe* (Belgium, Brussels, 2012), pp. 16–19
7. F.G. Bass, I.M. Fuks, *Scattering of Waves by a Statistically Rough Surface* (Moscow, 1972), p. 424

8. P. Beckmann, *The Scattering of Electromagnetic Waves from Rough Surfaces* (NY, 1963)
9. S.M. Rytov, *Introduction to Statistical Radiophysics* (Moscow, 1966), p. 404
10. G.V. Simonenko, V.V. Tuchin, N.A. Akodina, Measurement of the optical anisotropy of biological tissues using a cell with a nematic liquid crystal. J. Opt. Technol. **67**(6), 70–73 (2000)
11. S. Nickell, M. Hermann, Essenpreis anisotropy of light propagation in human skin. Phys. Med. Biol. **45**, 2873–2886 (2000)
12. A.V. Laktyunkin, Modelling of scattering of millimeter and centimeter wave fractal surfaces at low angles of incidence. Ph.D. dissertation, Moscow, 2009
13. A.A. Potapov, *Fractals in Radiophysics and Radiolocation* (M, 2002), p. 664

Chapter 6
Light Scattering by Dielectric Bodies of Irregular Shape in a Layered Medium

Abstract The mathematical model proposed for predicting optical characteristics (refractive index and absorption coefficient) of a biotissue being simulated, which is probed in vivo by a laser beam. Blood corpuscles in this case are simulated by particles of irregular shape and various sizes, which are oriented arbitrarily in free space. Using the mathematical model constructed earlier, optical characteristics (refractive index and absorption coefficient) of the biotissue being simulated, probed in vivo by a laser beam, are analyzed. The action spectra of the laser radiation power absorbed by oxyhemoglobin and deoxihemoglobin of blood, which are associated with selectivity of absorption of radiation by these hemoglobin derivatives, are calculated.

6.1 Introduction

In this chapter, we analyze optical and geometrical characteristics of particles simulating erythrocytes in the upper layer of the dermis. It should be noted that hemorheological and microcirculation disorders occupy the leading place in the pathogenesis of many diseases as well as states and complications. Functional properties of erythrocytes that form the major part of blood cells play the leading role in the formation of such pathological states. The erythrocyte as a physical object is characterized by the geometrical size, refractive index, and mechanical properties such as elasticity and deformability. For this reason, analysis of various characteristics of erythrocytes (in particular, their size, shape, and refractive index) in connection with various diseases of the blood system is of certain theoretical and undoubtedly of practical importance.

The possibility of theoretical construction of optical characteristics of dielectric particles of various shapes and structures has been studied in a number of publications [1–4].

We will solve the problem of scattering from bodies of an arbitrary shape by the method of integral equations, which is known as the extended boundary conditions method (EBCM) [5, 6]. It should be noted that this method gives the exact solution to the problem of light scattering by a particle of an arbitrary shape in the form of infinite series; however, the maximal number of expansion terms required for attaining a reasonable accuracy depends on the shape, size, and refractive index of

© Springer International Publishing AG, part of Springer Nature 2018   103
K. Kulikov and T. Koshlan, *Laser Interaction with Heterogeneous Biological Tissue*, Biological and Medical Physics, Biomedical Engineering, https://doi.org/10.1007/978-3-319-94114-1_6

the scatterer. We have constructed a mathematical model that makes it possible to vary the electrophysical and geometrical parameters (thickness of the layers) of the biological structure being simulated and to represent the result in the form of a graph describing the dependence of the laser radiation intensity on the electrophysical characteristics of the model structure for each version being analyzed.

The problem consists of several steps. At the first step, it is necessary to find the coefficient of reflection of a plane wave from a smoothly irregular layer simulating a given biological structure which consist of two continuous layers and the third layer containing inhomogeneous inclusions simulating blood cells with different refractive indices.

At the second step, it is necessary to solve the problem of reflection of a Gaussian beam with an arbitrary cross section for the above conditions (see Chap. 4). The construction of these parts is auxiliary. In this study, we precisely solve the problem of light scattering from particles of irregular shapes, which simulate erythrocytes oriented arbitrarily in free space, taking into account their multiple scattering, as well as the problem of simulating the efficiency of light absorption by the main derivatives of hemoglobin: oxyhemoglobin (HbO_2) and deoxyhemoglobin (Hb) of human blood in the upper layers of human dermis.

Chapter is based on the results of the [7–9].

6.2   Matrix Formulation of Scattering for the jth Particle of an Arbitrary Shape

From the standpoint of biomedical optics, whole blood is a highly concentrated turbid medium, whose scattering and absorption properties are mainly determined by erythrocytes. For this reason, we will consider in this section erythrocytes present in blood and their optical properties, disregarding the effect of other blood corpuscles on light scattering; this does not affect the generality and correctness of the formulation of the problem.

In some publications, an erythrocyte is considered as a homogeneous sphere whose volume is the same as that of the erythrocyte [10, 11]; this can be treated as the first approximation (see Chap. 3), and it is expedient to consider the erythrocyte as a body of an irregular shape.

Let us suppose that a plane linearly polarized electromagnetic wave is incident on a group of homogeneous particles simulating erythrocytes with radii a^j and refractive indices N^j, where j are the numbers of particles. The direction of the incident wave is arbitrary. The group of particles is considered in the 3D system of coordinates with the origin at the center of a particle with certain number j_0. We denote by $\mathbf{r}_{j_0, j}$ the radius vector of any other jth particle. We always assume that the surface (denoted by s) of a particle is quite regular and satisfies the Green theorem, and surface s of the scatterer has a continuous single-valued normal n at each point. We consider only the simple harmonic time dependence with circular frequency ω, omitting factor $\exp(-i\omega t)$

everywhere. We assume that the size of a particle simulating the erythrocyte is larger than the wavelength of incident radiation; i.e., $ka^j > 1$, where a^j is the radius of the jth particle simulating the erythrocyte.

We write the system of Maxwell equations for the field in the vicinity of the j_0th particle, which is distorted by other particles:

$$\nabla \times \mathbf{H} = -ik\varepsilon\mathbf{E}, \ \nabla \times \mathbf{E} = ik\mu\mathbf{H}, \quad \mathrm{div}\mathbf{E} = 0, \quad \mathrm{div}\mathbf{H} = 0.$$

At the boundary of the particle with the surrounding medium, we must impose the boundary conditions

$$\mathbf{n} \times \mathbf{E}_i - \mathbf{n} \times \mathbf{E}_s = \mathbf{n} \times \mathbf{E}_I, \quad \mathbf{n} \times \mathbf{H}_i - \mathbf{n} \times \mathbf{H}_s = \mathbf{n} \times \mathbf{H}_I, \tag{6.1}$$

where k is the wavenumber, ε is the permittivity of the medium, μ is the permeability of the medium, $\mathbf{E}_s$ is the scattered field, $\mathbf{E}_I$ is the incident field, and $\mathbf{E}_i$ is the internal field. The expressions for these fields will be given below.

The total field can be written in the form $\mathbf{E}(r') = \mathbf{E}_I(r') + \mathbf{E}_s(r')$. According to [12], we can write the corresponding integral equation

$$\mathbf{E}_I(r') + \nabla \times \int_s \mathbf{n} \times E(r)G(r, r')ds + \frac{i}{k\varepsilon}\nabla \times \nabla \times \int_s \mathbf{n} \times \mathbf{H}(r) \times$$

$$\times \, G(r, r')ds = 0, \tag{6.2}$$

where $G(r, r')$ is the Green function defined as [12]:

$$G(r, r') = \frac{ik}{\pi} \sum_{n=1}^{\infty} \sum_{m=-n}^{n} (-1)^m E_{mn}[\mathbf{M}^3_{-mn}(kr, \theta, \varphi) \cdot \mathbf{M}^1_{mn}(kr', \theta', \varphi') +$$

$$+ \, \mathbf{N}^3_{-mn}(kr, \theta, \varphi)\mathbf{N}^1_{mn}(kr', \theta', \varphi')], \tag{6.3}$$

for $r > r'$ and

$$G(r, r') = \frac{ik}{\pi} \sum_{n=1}^{\infty} \sum_{m=-n}^{n} (-1)^m E_{mn}[\mathbf{M}^1_{-mn}(kr, \theta, \varphi) \cdot \mathbf{M}^3_{mn}(kr', \theta', \varphi') +$$

$$+ \, \mathbf{N}^1_{-mn}(kr, \theta, \varphi)\mathbf{N}^3_{mn}(kr', \theta', \varphi')], \tag{6.4}$$

for $r' > r$, where $\mathbf{M}_{mn}, \mathbf{N}_{mn}, \mathbf{M}_{-mn}, \mathbf{N}_{-mn}$ are vector spherical harmonics.

Note that the vector spherical harmonics should be chosen on the basis of invariance (in the sense of closeness) property; namely, under the rotation of the system of coordinates, vector spherical harmonics $\mathbf{M}_{mn}$ and $\mathbf{N}_{mn}$ should be transformed independently.

The required properties of invariance are satisfied by the following spherical harmonics [12]:

$$\mathbf{M}_{mn}^{J}(kr) = (-1)^m d_n z_n^J(kr) \mathbf{C}_{mn}(\theta) \exp(im\varphi), \tag{6.5}$$

$$\mathbf{N}_{mn}^{J}(kr) = (-1)^m d_n \left[\frac{n(n+1)}{kr} z_n^J(kr) \mathbf{P}_{mn}(\theta) + \frac{1}{kr} z_n^J(kr) \mathbf{B}_{mn}(\theta) \right] \times$$

$$\times \exp(im\varphi), \tag{6.6}$$

$$\mathbf{B}_{mn}(\theta) = \mathbf{i}_\theta \frac{d}{d\theta} d_{om}^n(\theta) + \mathbf{i}_\varphi \frac{im}{\sin(\theta)} d_{om}^n(\theta), \tag{6.7}$$

$$\mathbf{C}_{mn}(\theta) = \mathbf{i}_\theta \frac{im}{\sin(\theta)} d_{om}^n(\theta) - \mathbf{i}_\varphi \frac{d}{d\theta} d_{om}^n(\theta), \tag{6.8}$$

$$\mathbf{P}_{mn}(\theta) = \mathbf{i}_r d_{om}^n(\theta), \ d_n = \sqrt{\frac{(2n+1)}{4n(n+1)}}, \tag{6.9}$$

where z_n^J is any of four spherical functions form (3.4),

$$d_{om}^n(\theta) = \frac{(-1)^{n-m}}{2^n n!} \left[\frac{(n+m)!}{(n-m)!} \right]^{1/2} (1 - \cos^2(\theta))^{-m/2} \times$$

$$\times \frac{d^{n-m}}{(d\cos(\theta))^{n-m}} [(1 - \cos^2(\theta)^n].$$

Let us write the expansion of the incident wave on the surface of the jth particle in the vector spherical harmonics:

$$\mathbf{E}_I(j) = -\sum_{n=1}^{\infty} \sum_{m=-n}^{n} i E_{mn} [p_{mn}^j \mathbf{N}_{mn}^1 + q_{mn}^j \mathbf{M}_{mn}^1]. \tag{6.10}$$

The expression for the internal field at the jth particle in vector spherical harmonics is:

$$\mathbf{E}_i(j) = -\sum_{n=1}^{\infty} \sum_{m=-n}^{n} i E_{mn} [d_{mn}^j \mathbf{N}_{mn}^1 + c_{mn}^j \mathbf{M}_{mn}^1], \tag{6.11}$$

The expansion for the field scattered by the jth particle in vector spherical harmonics has the form

$$\mathbf{E}_s(j) = \sum_{n=1}^{\infty} \sum_{m=-n}^{n} i E_{mn} [a_{mn}^j \mathbf{N}_{mn}^3 + b_{mn}^j \mathbf{M}_{mn}^3], \tag{6.12}$$

Following [12], we substitute expansions (6.10), (6.11) and (6.12) with allowance for expressions (6.3), (6.4), and boundary conditions (6.1) into integral equation (6.2); this gives

$$
\frac{ik^2}{\pi} \int_s \sum_{n=1}^{\infty} \sum_{m=-n}^{n} (-1)^m [c_{mn}^j \mathbf{n} \times \mathbf{M}_{m'n'}^1 + d_{mn}^j \mathbf{n} \times \mathbf{N}_{m'n'}^1] \begin{pmatrix} \mathbf{N}_{-mn}^3 \\ \mathbf{M}_{-mn}^3 \end{pmatrix} ds +
$$

$$
+ \frac{ik^2}{\pi} \sqrt{\frac{\varepsilon_1}{\mu_1}} \int_s \sum_{n=1}^{\infty} \sum_{m=-n}^{n} (-1)^m [c_{mn}^j \mathbf{n} \times \mathbf{N}_{m'n'}^1 + d_{mn}^j \mathbf{n} \times \mathbf{M}_{m'n'}^1] \times
$$

$$
\times \begin{pmatrix} \mathbf{M}_{-mn}^3 \\ \mathbf{N}_{-mn}^3 \end{pmatrix} ds = - \begin{pmatrix} p_{mn}^j \\ q_{mn}^j \end{pmatrix}
$$

or, in matrix form,

$$
\begin{pmatrix} I_1^{21} + \tilde{m} \cdot I_1^{12} & I_1^{22} + \tilde{m} \cdot I_1^{11} \\ I_1^{22} + \tilde{m} \cdot I_1^{11} & I_1^{12} + \tilde{m} \cdot I_1^{21} \end{pmatrix} \begin{pmatrix} d^j \\ c^j \end{pmatrix} = -i \begin{pmatrix} p^j \\ q^j \end{pmatrix}, \tag{6.13}
$$

where $\tilde{m}$ is the relative refraction index of the particle and

$$
\frac{ik^2}{\pi} \int_s \sum_{n=1}^{\infty} \sum_{m=-n}^{n} (-1)^m [c_{mn}^j \mathbf{n} \times \mathbf{M}_{m'n'}^1 + d_{mn}^j \mathbf{n} \times \mathbf{N}_{m'n'}^1] \begin{pmatrix} \mathbf{N}_{-mn}^1 \\ \mathbf{M}_{-mn}^1 \end{pmatrix} ds +
$$

$$
+ \frac{ik^2}{\pi} \sqrt{\frac{\varepsilon_1}{\mu_1}} \int_s \sum_{n=1}^{\infty} \sum_{m=-n}^{n} (-1)^m [c_{mn}^j \mathbf{n} \times \mathbf{N}_{m'n'}^1 + d_{mn}^j \mathbf{n} \times \mathbf{M}_{m'n'}^1] \times
$$

$$
\times \begin{pmatrix} \mathbf{M}_{-mn}^1 \\ \mathbf{N}_{-mn}^1 \end{pmatrix} ds = - \begin{pmatrix} a_{mn}^j \\ b_{mn}^j \end{pmatrix}
$$

or, in matrix form,

$$
\begin{pmatrix} a^j \\ b^j \end{pmatrix} = -i \begin{pmatrix} I_1'^{21} + \tilde{m} \cdot I_1'^{12} & I_1'^{22} + \tilde{m} \cdot I_1'^{11} \\ I_1'^{22} + \tilde{m} \cdot I_1'^{11} & I_1'^{12} + \tilde{m} \cdot I_1'^{21} \end{pmatrix} \begin{pmatrix} d^j \\ c^j \end{pmatrix}. \tag{6.14}
$$

Combining expressions (6.13) and (6.14), we obtain

$$
\begin{pmatrix} a^j \\ b^j \end{pmatrix} = - \begin{pmatrix} I_1'^{21} + \tilde{m} \cdot I_1'^{12} & I_1'^{22} + \tilde{m} \cdot I_1'^{11} \\ I_1'^{22} + \tilde{m} \cdot I_1'^{11} & I_1'^{12} + \tilde{m} \cdot I_1'^{21} \end{pmatrix} \begin{pmatrix} I_1^{21} + \tilde{m} \cdot I_1^{12} & I_1^{22} + \tilde{m} \cdot I_1^{11} \\ I_1^{22} + \tilde{m} \cdot I_1^{11} & I_1^{12} + \tilde{m} \cdot I_1^{21} \end{pmatrix}^{-1} \times
$$

$$
(\times) \begin{pmatrix} p^j \\ q^j \end{pmatrix}. \tag{6.15}
$$

Denoting matrices by Q_{01}^{11} and Q_{01}^{31} we can write expression (6.15) in the form

$$\begin{pmatrix} a^j \\ b^j \end{pmatrix} = T_1^j \begin{pmatrix} p^j \\ q^j \end{pmatrix}, T_1^j = -Q_{01}^{11}(k, k_1) \cdot [Q_{01}^{31}(k, k_1)]^{-1}. \tag{6.16}$$

The elements of matrix T_1^j can be expressed in the form of surface integrals:

$$I_{1mnm'n'}^{11} = \alpha(-1)^m \int_s [\mathbf{M}_{(-mn)}^3(kr) \times \mathbf{M}_{(m'n')}^1(k_1r)]\mathbf{n}dS, \tag{6.17}$$

$$I_{1mnm'n'}^{12} = \alpha(-1)^m \int_s [\mathbf{M}_{(-mn)}^3(kr) \times \mathbf{N}_{(m'n')}^1(k_1r)]\mathbf{n}dS, \tag{6.18}$$

$$I_{1mnm'n'}^{21} = \alpha(-1)^m \int_s [\mathbf{N}_{(-mn)}^3(kr) \times \mathbf{M}^1_{(m'n')}(k_1r)]\mathbf{n}dS, \tag{6.19}$$

$$I_{1mnm'n'}^{22} = \alpha(-1)^m \int_s [\mathbf{N}_{(-mn)}^3(kr) \times \mathbf{N}^1_{(m'n')}(k_1r)]\mathbf{n}dS, \tag{6.20}$$

$$I_{1mnm'n'}^{'11} = \alpha(-1)^{m'} \int_s [\mathbf{M}_{(-mn)}^1(kr) \times \mathbf{M}_{(m'n')}^1(k_1r)]\mathbf{n}dS, \tag{6.21}$$

$$I_{1mnm'n'}^{'12} = \alpha(-1)^{m'} \int_s [\mathbf{M}_{(-mn)}^1(kr) \times \mathbf{N}_{(m'n')}^1(k_1r)]\mathbf{n}dS, \tag{6.22}$$

$$I_{1mnm'n'}^{'21} = \alpha(-1)^{m'} \int_s [\mathbf{N}_{(-mn)}^1(kr) \times \mathbf{M}_{(m'n')}^1(k_1r)]\mathbf{n}dS, \tag{6.23}$$

$$I_{1mnm'n'}^{'22} = \alpha(-1)^{m'} \int_s [\mathbf{N}_{(-mn)}^1(kr) \times \mathbf{N}_{(m'n')}^1(k_1r)]\mathbf{n}dS, \tag{6.24}$$

where $\alpha = k^2/\pi$.

6.3 Explicit Expressions for the Integrals with Vector Products of Vector Spherical Functions

Let us write the expression for the normal of the object in the Cartesian system of coordinates:

$$\mathbf{n} = n_x\mathbf{i} + n_y\mathbf{j} + n_z\mathbf{k},$$

where $\mathbf{i}$, $\mathbf{j}$ and $\mathbf{k}$ are the unit vectors of the corresponding system. According to [13], for an arbitrarily oriented body, we have

$$\mathbf{n}dS = \left[\frac{\partial(y, z)}{\partial(\theta, \varphi)}\mathbf{i} + \frac{\partial(z, x)}{\partial(\theta, \varphi)}\mathbf{j} + \frac{\partial(x, y)}{\partial(\theta, \varphi)}\mathbf{k} \right],$$

whence

$$n_x dS = \left[rr'_\varphi \sin(\varphi) + r^2 \sin^2(\theta) \cos(\varphi) - rr'_\theta \sin(\theta) \cos(\varphi) \right] d\theta d\varphi,$$

$$n_y dS = \left[-rr'_\varphi \cos(\varphi) + r^2 \sin^2(\theta) \sin(\varphi) - rr'_\theta \sin(\theta) \sin(\varphi) \right] d\theta d\varphi,$$

$$n_z dS = \left[r^2 \sin^2(\theta) \sin(\theta) \cos(\theta) - rr'_\theta \sin^2(\theta) \right] d\theta d\varphi.$$

Using the formulas for transformation from the Cartesian coordinates into spherical coordinates, we obtain

$$n_r dS = \left[\sin(\theta) \cos(\varphi)n_x + \sin(\theta) \sin(\varphi)n_y + \cos(\theta)n_z \right] d\theta d\varphi = r^2 \sin(\theta) d\theta d\varphi,$$

$$n_\theta dS = \left[\cos(\theta) \cos(\varphi)n_x + \cos(\theta) \sin(\varphi)n_y - \sin(\theta)n_z \right] d\theta d\varphi = -rr'_\theta \sin(\theta) d\theta d\varphi,$$

$$n_\varphi dS = \left[- \sin(\varphi)n_x + \cos(\varphi)n_y \right] d\theta d\varphi = -rr'_\varphi d\theta d\varphi.$$

Substituting the expression for $\mathbf{n}dS$, $\mathbf{N}^1_{mn}$, $\mathbf{M}^1_{mn}$, $\mathbf{N}^3_{mn}$ and $\mathbf{M}^3_{mn}$ into surface integrals (6.17)–(6.24), we obtain

$$I^{11}_{mnm'n'} = (-1)^{(m+m')} \int_0^\pi i(md^n_{om}(\theta)b^{n'}_{om'}(\theta) + m'd^n_{om}(\theta)b^{n'}_{om'}(\theta)),$$

$$\left[\int_0^{2\pi} c^1_{mnm'n'}(\theta, \phi)d\varphi \right] d\theta \tag{6.25}$$

$$I^{12}_{mnm'n'} = (-1)^{(m+m')} \int_0^\pi -(b^n_{om}(\theta)b^{n'}_{om'}(\theta) \sin(\theta) + mm'd^n_{om}(\theta)d^{n'}_{om'}(\theta)/ \sin(\theta)) \times$$

$$\times \left[\int_0^{2\pi} c^2_{mnm'n'}(\theta, \phi)d\varphi \right] - \frac{n(n+1)}{x} d^n_{om}(\theta)b^{n'}_{om'}(\theta) \sin(\theta) \left[\int_0^{2\pi} c^3_{mnm'n'}(\theta, \phi)d\varphi \right] -$$

$$- i \frac{n'(n'+1)}{x_1} d^n_{om}(\theta)d^{n'}_{om'}(\theta) \sin(\theta) \left[\int_0^{2\pi} c^4_{mnm'n'}(\theta, \phi)d\varphi \right] d\theta, \tag{6.26}$$

$$I_{mnm'n'}^{21} = (-1)^{(m+m')} \int_0^\pi \frac{n'(n'+1)}{x_1} b_{om}^n(\theta) d_{om'}^{n'}(\theta) \sin(\theta) \left[\int_0^{2\pi} c_{mnm'n'}^3(\theta, \phi) d\varphi \right] -$$

$$-im \frac{n'(n'+1)}{x_1} d_{om}^n(\theta) d_{om'}^{n'}(\theta) / \sin(\theta) \left[\int_0^{2\pi} c_{mnm'n'}^4(\theta, \phi) d\varphi \right] +$$

$$+ (mm' d_{om}^n(\theta) d_{om'}^{n'}(\theta) / \sin(\theta) +$$

$$+ b_{om}^n(\theta) b_{om'}^{n'}(\theta) \sin(\theta) \left[\int_0^{2\pi} c_{mnm'n'}^5(\theta, \phi) d\varphi \right] d\theta, \qquad (6.27)$$

$$I_{mnm'n'}^{22} = (-1)^{(m+m')} \int_0^\pi i(m d_{om}^n(\theta) b_{om'}^{n'}(\theta) + m' d_{om'}^{n'}(\theta) b_{om}^n(\theta)) \left[\int_0^{2\pi} c_{mnm'n'}^6(\theta, \phi) d\varphi \right] +$$

$$+ \frac{n'(n'+1)}{x_1} b_{om}^n(\theta) d_{om'}^{n'}(\theta) \left[\int_0^{2\pi} c_{mnm'n'}^7(\theta, \phi) d\varphi \right] - \frac{n(n+1)}{x} d_{om}^n(\theta) b_{om'}^{n'}(\theta) \times$$

$$\times \left[\int_0^{2\pi} c_{mnm'n'}^8(\theta, \phi) d\varphi \right] + im \frac{n'(n'+1)}{x_1} d_{om}^n(\theta) d_{om'}^{n'}(\theta) \left[\int_0^{2\pi} c_{mnm'n'}^9(\theta, \phi) d\varphi \right] +$$

$$+ im' \frac{n(n+1)}{x} d_{om}^n(\theta) d_{om'}^{n'}(\theta) \left[\int_0^{2\pi} c_{mnm'n'}^{10}(\theta, \phi) d\varphi \right] d\theta, \qquad (6.28)$$

$$I_{mnm'n'}^{'11} = (-1)^{(m+m')} \int_0^\pi i(m d_{om}^n(\theta) b_{om'}^{n'}(\theta) + m' d_m^n(\theta) b_{om'}^{n'}(\theta)) \times$$

$$\times \left[\int_0^{2\pi} f_{mnm'n'}^1(\theta, \phi) d\varphi \right] d\theta \qquad (6.29)$$

$$I_{mnm'n'}^{'12} = (-1)^{(m+m')} \int_0^\pi -(b_{om}^n(\theta) b_{om'}^{n'}(\theta) \sin(\theta) + mm' d_{om}^n(\theta) d_{om'}^{n'}(\theta) / \sin(\theta)) \times$$

$$\times \left[\int_0^{2\pi} f_{mnm'n'}^2(\theta, \phi) d\varphi \right] - \frac{n(n+1)}{x} d_{om}^n(\theta) b_{om'}^{n'}(\theta) \sin(\theta) \left[\int_0^{2\pi} f_{mnm'n'}^3(\theta, \phi) d\varphi \right] -$$

$$- i \frac{n'(n'+1)}{x_1} d_{om}^n(\theta) d_{om'}^{n'}(\theta) \sin(\theta) \left[\int_0^{2\pi} f_{mnm'n'}^4(\theta, \phi) d\varphi \right] d\theta, \qquad (6.30)$$

$$I'^{21}_{mnm'n'} = (-1)^{(m+m')} \int_0^\pi \frac{n'(n'+1)}{x_1} b^n_{om}(\theta) d^{n'}_{om'}(\theta) \sin(\theta) \left[\int_0^{2\pi} f^3_{mnm'n'}(\theta, \phi) d\varphi \right] -$$

$$-im \frac{n'(n'+1)}{x_1} d^n_{om}(\theta) d^{n'}_{om'}(\theta)/\sin(\theta) \left[\int_0^{2\pi} f^4_{mnm'n'}(\theta, \phi) d\varphi \right] +$$

$$+ (mm' d^n_{om}(\theta) d^{n'}_{om'}(\theta)/\sin(\theta) + \tag{6.31}$$

$$+ b^n_{om}(\theta) b^{n'}_{om'}(\theta) \sin(\theta) \left[\int_0^{2\pi} f^5_{mnm'n'}(\theta, \phi) d\varphi \right] d\theta,$$

$$I'^{22}_{mnm'n'} = (-1)^{(m+m')} \int_0^\pi i(md^n_{om}(\theta) b^{n'}_{om'}(\theta) + m' d^{n'}_{om'}(\theta) b^n_{om}(\theta)) \left[\int_0^{2\pi} f^6_{mnm'n'}(\theta, \phi) d\varphi \right] +$$

$$+ \frac{n'(n'+1)}{x_1} b^n_{om}(\theta) d^{n'}_{om'}(\theta) \left[\int_0^{2\pi} f^7_{mnm'n'}(\theta, \phi) d\varphi \right] - \frac{n(n+1)}{x} d^n_{om}(\theta) b^{n'}_{om'}(\theta) \times$$

$$\times \left[\int_0^{2\pi} f^8_{mnm'n'}(\theta, \phi) d\varphi \right] + im \frac{n'(n'+1)}{x_1} d^n_{om}(\theta) d^{n'}_{om'}(\theta) \left[\int_0^{2\pi} f^9_{mnm'n'}(\theta, \phi) d\varphi \right] +$$

$$+ im' \frac{n(n+1)}{x} d^n_{om}(\theta) d^{n'}_{om'}(\theta) \left[\int_0^{2\pi} f^{10}_{mnm'n'}(\theta, \phi) d\varphi \right] d\theta, \tag{6.32}$$

where

$$c^1_{mnm'n'}(\theta, \phi) = \exp[i \triangle_{m'm}] h_n(x) j_n(x_1) r^2(\theta, \phi),$$

$$c^2_{mnm'n'}(\theta, \phi) = \exp[i \triangle_{m'm}] u_n(x) j_n(x_1) r^2(\theta, \phi),$$

$$c^3_{mnm'n'}(\theta, \phi) = \exp[i \triangle_{m'm}] h_n(x) j_n(x_1) \frac{dr(\theta, \phi)}{d\theta},$$

$$c^4_{mnm'n'}(\theta, \phi) = \exp[i \triangle_{m'm}] h_n(x) j_n(x_1) \frac{dr(\theta, \phi)}{d\theta},$$

$$c^5_{mnm'n'}(\theta, \phi) = \exp[i \triangle_{m'm}] h_n(x) v_n(x_1) r^2(\theta, \phi),$$

$$c^6_{mnm'n'}(\theta, \phi) = \exp[i \triangle_{m'm}] u_n(x) v_n(x_1) r^2(\theta, \phi),$$

$$c^7_{mnm'n'}(\theta, \phi) = \exp[i \triangle_{m'm}] u_n(x) j_n(x_1) \frac{dr(\theta, \phi)}{d\varphi},$$

$$c^8_{mnm'n'}(\theta, \phi) = \exp[i \triangle_{m'm}] h_n(x) v_n(x_1) \frac{dr(\theta, \phi)}{d\varphi},$$

$$c^9_{mnm'n'}(\theta, \phi) = \exp[i \triangle_{m'm}] u_n(x) j_n(x_1) \frac{dr(\theta, \phi)}{d\theta},$$

$$c^{10}_{mnm'n'}(\theta, \phi) = \exp[i \triangle_{m'm}] h_n(x) v_n(x_1) \frac{dr(\theta, \phi)}{d\theta},$$

$$f^1_{mnm'n'}(\theta, \phi) = \exp[i \triangle_{m'm}] j_n(x) j_n(x_1) r^2(\theta, \phi),$$

$$f^2_{mnm'n'}(\theta, \phi) = \exp[i \triangle_{m'm}] v_n(x) j_n(x_1) r^2(\theta, \phi),$$

$$f^3_{mnm'n'}(\theta, \phi) = \exp[i \triangle_{m'm}] j_n(x) j_n(x_1) \frac{dr(\theta, \phi)}{d\theta},$$

$$f^4_{mnm'n'}(\theta, \phi) = \exp[i \triangle_{m'm}] j_n(x) j_n(x_1) \frac{dr(\theta, \phi)}{d\theta},$$

$$f^5_{mnm'n'}(\theta, \phi) = \exp[i \triangle_{m'm}] j_n(x) v_n(x_1) r^2(\theta, \phi),$$

$$f^6_{mnm'n'}(\theta, \phi) = \exp[i \triangle_{m'm}] v_n(x) v_n(x_1) r^2(\theta, \phi),$$

$$f^7_{mnm'n'}(\theta, \phi) = \exp[i \triangle_{m'm}] v_n(x) j_n(x_1) \frac{dr(\theta, \phi)}{d\varphi},$$

$$f^8_{mnm'n'}(\theta, \phi) = \exp[i \triangle_{m'm}] j_n(x) v_n(x_1) \frac{dr(\theta, \phi)}{d\varphi},$$

$$f^9_{mnm'n'}(\theta, \phi) = \exp[i \triangle_{m'm}] v_n(x) j_n(x_1) \frac{dr(\theta, \phi)}{d\theta},$$

$$f^{10}_{mnm'n'}(\theta, \phi) = \exp[i \triangle_{m'm}] j_n(x) v_n(x_1) \frac{dr(\theta, \phi)}{d\theta},$$

where

$$u_n(x) = \frac{1}{x} \frac{d}{dx}(x h_n(x)), \quad v_n(x) = \frac{1}{x} \frac{d}{dx}(x j_n(x)),$$

$$\triangle_{m'm} = (m' - m), \quad b^n_{om}(\theta) = \frac{d}{d\theta} d^n_{om}(\theta), \quad x = k \cdot r(\theta, \phi)$$

$x_1 = \tilde{m} \cdot k \cdot r(\theta, \phi)$, $k = 2\pi/\lambda$, λ is the wavelength of incident radiation and $r(\theta, \phi)$ is the equation of the particle surface in the spherical system of coordinates.

6.4 Matrix Formulation of Scattering for the jth Bilayer Particle of an Arbitrary Shape

Specific properties of biological particles (blood corpuscles) require a more sophisticated and more adequate model due to the existence of a core and a plasma membrane in the object under investigation.

Let r_1 be the radius of the cell core and r_2 be the radius of the plasma membrane. We consider scattering of a plane electromagnetic wave by the jth inhomogeneous particle of irregular shape (see Fig. 6.1). Surface S_1 is defined in coordinate system $O_1x_1y_1z_1$, while surface S_2 is defined in coordinate system $O_2x_2y_2z_2$.

We write the system of Maxwell equations for the corresponding fields:

$$\nabla \times \mathbf{H_s} = -ik\varepsilon\mathbf{E_s}, \nabla \times \mathbf{E_s} = ik\mu\mathbf{H_s}$$

for domain D,

$$\nabla \times \mathbf{H_1} = -ik\varepsilon_1\mathbf{E_1} \nabla \times \mathbf{E_1} = ik\mu_1\mathbf{H_1}$$

for domain S_1, and

$$\nabla \times \mathbf{H_2} = -ik\varepsilon_2\mathbf{E_2}, \nabla \times \mathbf{E_2} = ik\mu_2\mathbf{H_2}$$

for domain S_2.

These fields must satisfy the following boundary conditions:

$$\mathbf{n_2} \times \mathbf{E_1} = \mathbf{n_2} \times \mathbf{E_2}, \mathbf{n_2} \times \mathbf{H_1} = \mathbf{n_2} \times \mathbf{H_2}$$

for domain S_1 and

$$\mathbf{n_1} \times \mathbf{E_1} - \mathbf{n_1} \times \mathbf{E_s} = \mathbf{n_1} \times \mathbf{E_I}, \quad \mathbf{n_1} \times \mathbf{H_1} - \mathbf{n_1} \times \mathbf{H_s} = \mathbf{n_1} \times \mathbf{H_I},$$

Fig. 6.1 The geometry of heterogeneous particles

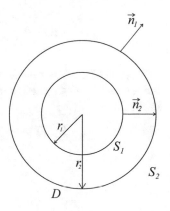

for domain S_2, where k is the wavenumber, ε is the permittivity of the medium, μ is the permeability of the medium, ε_1 is the permittivity of the cell core, ε_2 is the permittivity of the plasma membrane, $\mathbf{E}_s$ is the scattered field, $\mathbf{E}_I$ is the incident field, $\mathbf{E}_1$ is the internal field, and $\mathbf{E}_2$ will be defined below.

Let us write the following integral equations [14]:

$$
\mathbf{E}_I(r_1) + \nabla \times \int_{S_2} \mathbf{n_1} \times \mathbf{E}_1(r_1')G(r_1, r_1')ds(r_1') + \frac{i}{k\varepsilon}\nabla \times \nabla \times \int_{S_2} \mathbf{n_1} \times \mathbf{H}_1(r_1') \times
$$

$$
\times G(r_1, r_1')ds(r_1') = 0, \tag{6.33}
$$

$$
-\nabla \times \int_{S_2} [\mathbf{n_2} \times \mathbf{E}_2(r_1')]G(r_1, r_1')ds(r_1') -
$$

$$
-\frac{i}{k\varepsilon_2}\nabla \times \nabla \times \int_{S_2} [\mathbf{n_2} \times \mathbf{H}_2(r_1')]G(r_1, r_1')ds(r_1') +
$$

$$
+\nabla \times \int_{S_1} [\mathbf{n_1} \times \mathbf{E}_1(r_1'')]G(r_1, r_1'')ds(r_1'') + \frac{i}{k\varepsilon_1}\nabla \times \nabla \times \int_{S_1} [\mathbf{n_1} \times \mathbf{H}_1(r_1'')] \times
$$

$$
\times G(r_1, r_1'')ds(r_1'') = 0,
$$

where $G(r, r')$ and $G(r, r'')$ are the Green functions.

The expression for the field scattered by the jth particle has the form

$$
\mathbf{E}_s(j) = \sum_{n=1}^{\infty} \sum_{m=-n}^{n} i E_{mn}[a_{mn}^j \mathbf{N}_{mn}^3 + b_{mn}^j \mathbf{M}_{mn}^3], \tag{6.34}
$$

Let us write the expansion of the wave incident on the surface of the jth particle in vector spherical harmonics

$$
\mathbf{E}_I(j) = -\sum_{n=1}^{\infty} \sum_{m=-n}^{n} i E_{mn}[p_{mn}^j \mathbf{N}_{mn}^1 + q_{mn}^j \mathbf{M}_{mn}^1]. \tag{6.35}
$$

In view of the finiteness of the field at the center, the internal field of the particle in the region $0 \leq r \leq r_1$(i.e., in the vicinity of the center of the particle) can be written in the form

$$
\mathbf{E}_1(j) = -\sum_{n=1}^{\infty} \sum_{m=-n}^{n} i E_{mn}[d_{mn}^j \mathbf{N}_{mn}^1 + c_{mn}^j \mathbf{M}_{mn}^1], \tag{6.36}
$$

In the region $r_1 \leq r \leq r_2$, the internal field can be written as [14, 15]

$$\mathbf{E}_2(j) = -\sum_{n=1}^{\infty} \sum_{m=-n}^{n} i\,E_{mn}[\alpha_{mn}^{j}\mathbf{N}_{mn}^{1} + \beta_{mn}^{j}\mathbf{M}_{mn}^{1} +$$

$$+ \gamma\mathbf{N}_{mn}^{3} + \delta_{mn}^{j}\mathbf{M}_{mn}^{3}] \tag{6.37}$$

Proceeding analogously to the case of scattering from a homogeneous particle of an irregular shape, we obtain the solution to the scattering problem for a bilayer particle of an arbitrary geometry:

$$\begin{pmatrix} a^j \\ b^j \end{pmatrix} = T_2^{j} \begin{pmatrix} p^j \\ q^j \end{pmatrix}, \tag{6.38}$$

$T_2^{j} = -[Q_2^{11}(k, k_2) + Q_2^{13}(k, k_2)] \cdot D[[Q_2^{31}(k, k_2) + Q_2^{33}(k, k_2)] \cdot D]^{-1}$, where

$$D = S_{12} \cdot T_{01}^{j} \cdot S_{21}, \quad T_{01}^{j} = -Q_{01}^{11}(k_2, k_1) \cdot [Q_{01}^{31}(k_2, k_1)]^{-1}$$

$$(Q)_2^{13} = \begin{pmatrix} K^{13} + m_2/m \cdot J^{13} & I^{13} + m_2/m \cdot L^{13} \\ L^{13} + m_2/m \cdot I^{13} & J^{13} + m_2/m \cdot K^{13} \end{pmatrix}, \tag{6.39}$$

$$(Q)_2^{31} = \begin{pmatrix} K^{31} + m_2/m \cdot J^{31} & I^{31} + m_2/m \cdot L^{31} \\ L^{13} + m_2/m \cdot I^{31} & J^{31} + m_2/m \cdot K^{31} \end{pmatrix}, \tag{6.40}$$

$$(Q)_2^{11} = \begin{pmatrix} K^{11} + m_2/m \cdot J^{11} & I^{11} + m_2/m \cdot L^{11} \\ L^{11} + m_2/m \cdot I^{11} & J^{11} + m_2/m \cdot K^{11} \end{pmatrix}, \tag{6.41}$$

$$(Q)_2^{33} = \begin{pmatrix} K^{33} + m_2/m \cdot J^{33} & I^{33} + m_2/m \cdot L^{33} \\ L^{13} + m_2/m \cdot I^{33} & J^{33} + m_2/m \cdot K^{33} \end{pmatrix}, \tag{6.42}$$

$$(Q)_{01}^{11} = \begin{pmatrix} I_2'^{21} + m_1/m_2 \cdot I_2'^{12} & I_2'^{22} + m_1/m_2 \cdot I_2'^{11} \\ I_2'^{22} + m_1/m_2 \cdot I_2'^{11} & I_2'^{12} + m_1/m_2 \cdot I_2'^{21} \end{pmatrix}, \tag{6.43}$$

$$(Q)_{01}^{31} = \begin{pmatrix} I_2^{21} + m_1/m_2 \cdot I_2^{12} & I_2^{22} + m_1/m_2 \cdot I_2^{11} \\ I_2^{22} + m_1/m_2 \cdot I_2^{11} & I_2^{12} + m_1/m_2 \cdot I_2^{21} \end{pmatrix}, \tag{6.44}$$

m_1 is the refractive index of the core, m_2 is the refractive index of the plasma membrane, m is the refractive index of the medium and matrix elements Q_{01}^{31}, Q_{01}^{11}, Q_2^{13}, Q_2^{31}, Q_2^{11} and Q_2^{33} can be expressed in the form of surface integrals:

$$K_{mnm'n'}^{31} = \alpha(-1)^m \int_s [\mathbf{N}_{(-mn)}^{3}(kr) \times \mathbf{M}_{(m'n')}^{1}(k_2 r)]\mathbf{n}dS, \tag{6.45}$$

$$J^{31}_{\text{mnm}'n'} = \alpha(-1)^m \int_s [\mathbf{M}^3_{(-mn)}(kr) \times \mathbf{N}^1_{(m'n')}(k_2 r)]\mathbf{n}\mathrm{d}S, \qquad (6.46)$$

$$I^{31}_{\text{mnm}'n'} = \alpha(-1)^m \int_s [\mathbf{N}^3_{(-mn)}(kr) \times \mathbf{N}^1_{(m'n')}(k_2 r)]\mathbf{n}\mathrm{d}S, \qquad (6.47)$$

$$L^{31}_{\text{mnm}'n'} = \alpha(-1)^m \int_s [\mathbf{M}^3_{(-mn)}(kr) \times \mathbf{M}^1_{(m'n')}(k_2 r)]\mathbf{n}\mathrm{d}S, \qquad (6.48)$$

$$K^{13}_{\text{mnm}'n'} = \alpha(-1)^m \int_s [\mathbf{N}^1_{(-mn)}(kr) \times \mathbf{M}^3_{(m'n')}(k_2 r)]\mathbf{n}\mathrm{d}S, \qquad (6.49)$$

$$J^{13}_{\text{mnm}'n'} = \alpha(-1)^m \int_s [\mathbf{M}^1_{(-mn)}(kr) \times \mathbf{N}^3_{(m'n')}(k_2 r)]\mathbf{n}\mathrm{d}S, \qquad (6.50)$$

$$I^{13}_{\text{mnm}'n'} = \alpha(-1)^m \int_s [\mathbf{N}^1_{(-mn)}(kr) \times \mathbf{N}^3_{(m'n')}(k_2 r)]\mathbf{n}\mathrm{d}S, \qquad (6.51)$$

$$L^{13}_{\text{mnm}'n'} = \alpha(-1)^m \int_s [\mathbf{M}^1_{(-mn)}(kr) \times \mathbf{M}^3_{(m'n')}(k_2 r)]\mathbf{n}\mathrm{d}S, \qquad (6.52)$$

$$K^{11}_{\text{mnm}'n'} = \alpha(-1)^m \int_s [\mathbf{N}^1_{(-mn)}(kr) \times \mathbf{M}^1_{(m'n')}(k_2 r)]\mathbf{n}\mathrm{d}S, \qquad (6.53)$$

$$J^{11}_{\text{mnm}'n'} = \alpha(-1)^m \int_s [\mathbf{M}^1_{(-mn)}(kr) \times \mathbf{N}^1_{(m'n')}(k_2 r)]\mathbf{n}\mathrm{d}S, \qquad (6.54)$$

$$I^{11}_{\text{mnm}'n'} = \alpha(-1)^m \int_s [\mathbf{N}^1_{(-mn)}(kr) \times \mathbf{N}^1_{(m'n')}(k_2 r)]\mathbf{n}\mathrm{d}S, \qquad (6.55)$$

$$L^{11}_{\text{mnm}'n'} = \alpha(-1)^m \int_s [\mathbf{M}^1_{(-mn)}(kr) \times \mathbf{M}^1_{(m'n')}(k_2 r)]\mathbf{n}\mathrm{d}S, \qquad (6.56)$$

$$K^{33}_{\text{mnm}'n'} = \alpha(-1)^m \int_s [\mathbf{N}^3_{(-mn)}(kr) \times \mathbf{M}^3_{(m'n')}(k_2 r)]\mathbf{n}\mathrm{d}S, \qquad (6.57)$$

$$J^{33}_{\text{mnm}'n'} = \alpha(-1)^m \int_s [\mathbf{M}^3_{(-mn)}(kr) \times \mathbf{N}^3_{(m'n')}(k_2 r)]\mathbf{n}\mathrm{d}S, \qquad (6.58)$$

$$I^{33}_{\text{mnm}'n'} = \alpha(-1)^m \int_s [\mathbf{N}^3_{(-mn)}(kr) \times \mathbf{N}^3_{(m'n')}(k_2 r)]\mathbf{n}\mathrm{d}S, \qquad (6.59)$$

$$L^{33}_{\text{mnm}'n'} = \alpha(-1)^m \int_s [\mathbf{M}^3_{(-mn)}(kr) \times \mathbf{M}^3_{(m'n')}(k_2 r)]\mathbf{n}\mathrm{d}S, \qquad (6.60)$$

$$I^{11}_{2mnm'n'} = \alpha(-1)^m \int_s [\mathbf{M}^3_{(-mn)}(k_2 r) \times \mathbf{M}^1_{(m'n')}(k_1 r)]\mathbf{n}dS, \qquad (6.61)$$

$$I^{12}_{2mnm'n'} = \alpha(-1)^m \int_s [\mathbf{M}^3_{(-mn)}(k_2 r) \times \mathbf{N}^1_{(m'n')}(k_1 r)]\mathbf{n}dS, \qquad (6.62)$$

$$I^{21}_{2mnm'n'} = \alpha(-1)^m \int_s [\mathbf{N}^3_{(-mn)}(k_2 r) \times \mathbf{M}^1_{(m'n')}(k_1 r)]\mathbf{n}dS, \qquad (6.63)$$

$$I^{22}_{2mnm'n'} = \alpha(-1)^m \int_s [\mathbf{N}^3_{(-mn)}(k_2 r) \times \mathbf{N}^1_{(m'n')}(k_1 r)]\mathbf{n}dS, \qquad (6.64)$$

$$I'^{11}_{2mnm'n'} = \alpha(-1)^{m'} \int_s [\mathbf{M}^1_{(-mn)}(k_2 r) \times \mathbf{M}^1_{(m'n')}(k_1 r)]\mathbf{n}dS, \qquad (6.65)$$

$$I'^{12}_{2mnm'n'} = \alpha(-1)^{m'} \int_s [\mathbf{M}^1_{(-mn)}(k_2 r) \times \mathbf{N}^1_{(m'n')}(k_1 r)]\mathbf{n}dS, \qquad (6.66)$$

$$I'^{21}_{2mnm'n'} = \alpha(-1)^{m'} \int_s [\mathbf{N}^1_{(-mn)}(k_2 r) \times \mathbf{M}^1_{(m'n')}(k_1 r)]\mathbf{n}dS, \qquad (6.67)$$

$$I'^{22}_{2mnm'n'} = \alpha(-1)^{m'} \int_s [\mathbf{N}^1_{(-mn)}(k_2 r) \times \mathbf{N}^1_{(m'n')}(k_1 r)]\mathbf{n}dS, \qquad (6.68)$$

where $\alpha = k^2/\pi$, $S_{12} = \tau^{11} R(\alpha, \beta, \gamma)$, matrix S_{12} connects vector spherical waves defined in coordinate system $O_1 x_1 y_1 z_1$ with those defined in coordinate system $O_2 x_2 y_2 z_2$ (see Fig. 6.1) and can be expressed in the form of the product of the matrix of transfer from one coordinate system to the other and the rotation matrix $S_{21} = R(-\gamma, -\beta, -\alpha)\tau^{33}$ is the matrix that describes the inverse transformation, where $R(-\gamma, -\beta, -\alpha) = R^{-1}(\alpha, \beta, \gamma)$ and τ^{33}, τ^{11} are defined in [14],

$$R(\alpha, \beta, \gamma) = \begin{pmatrix} R_{mn,m'n'}(\alpha, \beta, \gamma) & 0 \\ 0 & R_{mn,m'n'}(\alpha, \beta, \gamma) \end{pmatrix},$$

$$R_{mn,m'n'}(\alpha, \beta, \gamma) = D^n_{mm'}(\alpha, \beta, \gamma)\delta_{nn'},$$

$$D^n_{mm'}(\alpha, \beta, \gamma) = (-1)^{(m+m')} \exp(im\alpha)d^{\sim m}_{mm'}(\beta)\exp(im'\gamma),$$

where D are Wigner functions, which are determined by the matrix elements of the irreducible representation of weight n on the rotation group [16] or as the matrix elements of the operator rotation $D(\alpha, \beta, \gamma)$ in the JM- representation :

$$<JM|D(\alpha, \beta, \gamma)|J'M'> = \delta_{JJ'} D^J_{mm'}(\alpha, \beta, \gamma).$$

Functions $D_{mm'}^n(\alpha, \beta, \gamma)$ can be represented as the product of three factors, each of which depends on only one Euler angle [17]

$$D_{mm'}^n(\alpha, \beta, \gamma) = \exp(-im\alpha)d_{mm'}^n(\beta)\exp(-im'\gamma),$$

where $d_{mm'}^n(\beta)$ are Wigner functions and satisfy the conditions of unitarity

$$\left[D^{-1}(\alpha, \beta, \gamma)\right]_{mm'}^n = \left[D^*(\alpha, \beta, \gamma)\right]_{mm'}^n,$$

$$\sum_{m=-n}^{n} D_{mm'}^*(\alpha, \beta, \gamma) D_{mm'}^{n*}(\alpha, \beta, \gamma) = \sum_{m=-n}^{n} D_{mm'}^n(\alpha, \beta, \gamma) D_{mm'}^{n-1}(\alpha, \beta, \gamma) = \delta_{m'm_1}$$

and orthogonality

$$\frac{2n+1}{8\pi^2} \int_0^{2\pi} d\alpha \int_0^{\pi} \sin\beta d\beta \int_0^{2\pi} d\gamma \, D_{mm'}^n(\alpha\beta\gamma) D_{m_1m_1'}^{n_1*}(\alpha\beta\gamma) = \delta_{nn'}\delta_{mm_1}\delta_{m'm_1'},$$

For function $d_{mm'}^n(\beta)$, we have

$$\int_0^{\pi} \sin\beta d\beta d_{mm'}^n(\beta) d_{mm'}^{n'}(\beta) = \frac{2}{2n+1}\delta_{nn'}.$$

Functions $d_{mm'}^n(\beta)$ satisfy the following relations

$$\frac{m}{\sin\beta}d_{om}^n(\beta)\Big|_{\beta=0}^{=1/2\delta_{m_1}[n(n+1)]^{1/2}}, \quad \frac{d}{d\beta}d_{om}^n(\beta)\Big|_{\beta=0}^{=1/2m\delta_{m_1}[n(n+1)]^{1/2}},$$

$$\frac{m}{\sin\beta}d_{om}^n(\beta) = 1/2[n(n+1)]^{1/2}[d_{1m}^n(\beta) + d_{-1m}^n(\beta)],$$

$$\frac{d}{d\beta}d_{om}^n(\beta) = 1/2[n(n+1)]^{1/2}[d_{1m}^n(\beta) - d_{-1m}^n(\beta)],$$

$$d_{mm'}^n(\beta)d_{m_1m_1'}(\beta) = \sum_{n_1=|n-n'|}^{n+n'} C_{nmn'm_1}^{n_1m+m_1} C_{nm'n'm_1'}^{n_1m'+m_1'} d_{m+m_1m'+m_1'}^{n_1}(\beta),$$

where $C_{nmn'm_1}^{n_1m+m_1}$, $C_{nm'n'm_1'}^{n_1m'+m_1'}$ are the Clebsch–Gordan coefficients

$$d_{mm'}^n(\beta) = (-1)^{m'-m}d_{-m-m'}^n(\beta) = (-1)^{m'-m}d_{m'm}^n(\beta).$$

The product of two D-functions $D_{m_1m_1'}^{m_1}(\alpha\beta\gamma) \, D_{m_2m_2'}^{n_2}(\alpha\beta\gamma)$ can be written as the following sum [17]:

$$D_{m_1m_1'}^{n_1}(\alpha\beta\gamma)D_{m_2m_2'}^{n_2}(\alpha\beta\gamma) = \sum_{n_3=|n_1-n_2|}^{n_1+n_2} C_{n_1m_1n_2m_2}^{n_3m_1+m_2} D_{m_1+m_2m_1'+m_2'}^{n_3}(\alpha\beta\gamma)C_{n_1m_1'n_2m_2'}^{n_3m_1'+m_2'}.$$

The formula of addition for D-functions of Wigner is [17]

$$\sum_{m^\sim=-n}^{n} D_{mm^\sim}^{n}(\alpha_1\beta_1\gamma_1)D_{m^\sim m'}^{n}(\alpha_2\beta_2\gamma_2) = D_{mm'}^{n}(\alpha\beta\gamma),$$

where $\alpha_1, \beta_1, \gamma_1$ are Euler's angles and characterize the rotation of the coordinate system $S \rightarrow S_1, \alpha_2, \beta_2, \gamma_2$–$S_1 \rightarrow S_2$, resulting rotation angles $S \rightarrow S_2$–α, β, γ relative to the original (S) coordinate system.

Recurrence relation for the calculation of the Wigner functions is

$$n\sqrt{(n+1)^2 - m^2}\sqrt{(n+1)^2 - q^2}d_{qm}^{n+1}(\beta)+$$

$$+(n+1)\sqrt{n^2 - q^2}\sqrt{n^2 - m^2}d_{qm}^{n-1}(\beta) = (2n+1(n(n+1)\cos\beta - mq)d_{qm}^{n}(\beta),$$

with initial conditions

$$d_{qm}^{n_*} = \frac{(-1)^{(q-m+|q-m|)}}{2^{n_*}}\left[\frac{(2_{n_*})!}{(|q-m|)!(|q+m|)!}\right]^{1/2}(1-\cos\beta)^{|q-m|/2}\times$$

$$\times(1+\cos\beta)^{|q+m|/2}, n_* - max(|m|,|q|).$$

Thus, the expansion coefficients of scattered and incident fields are connected by the linear transformation of the T-matrix that is invariant to the direction of propagation of incident radiation in a fixed system of coordinates and depends on the physical and geometrical characteristics of the scatterer (such as the refractive index, the size relative to the wavelength of light, and morphology). The above representation of the T- matrix method has certain advantages as compared to other representations, which lie in the use of vector spherical harmonics invariant to the rotation of the coordinate system and in the symmetric form of the representation of the main relations. It should be noted that the method of the T-matrix is a direct generalization of the standard Mie theory to the case of nonspherical particles. Indeed, if a scatterer is spherically symmetric, then the T-matrix becomes diagonal, and the diagonal elements are defined to within the sign by the relevant Mie coefficients a_n and b_n.

Note that the numerical calculation of the integrals with the vector products of the vector spherical functions for an arbitrary body of revolution is problematic in the case when the size of a scattering object is much larger than the wavelength of light. This is due to the fact that the integrand in the formula for computing elements of matrix $Q_{01}^{11}, Q_{01}^{31}, Q_2^{32}, Q_2^{33}, Q_2^{11}, Q_2^{13}$ may oscillate in very large limits, which leads to loss of accuracy. The process of numerical inversion of matrix Q_{01}^{11}, Q_{01}^{31}, Q_2^{32}, Q_2^{33} is poorly substantiated and also becomes unstable. Note that this is observed for particles with zero or a very small imaginary part of the refractive index.

It was shown in [18, 19] that effective approaches to improving the convergence of computations, which are based on the EBCM, are as follows.

1. Computation of elements of the matrix and its inversion using fourfold accuracy.

2. Inversion of the Q matrix using the LU factorization method. The electromagnetic field of the wave incident on the surface of the jth particle consists of two parts: the field of the initial wave and the field of the wave scattered by a group of other particles located in the surrounding medium. Then, we can write the following expression

$$\mathbf{E_i}(j) = \mathbf{E_0}(j) + \sum_{l \neq j} \mathbf{E_s}(l, j), \tag{6.69}$$

where $\mathbf{E_s}(l, j)$ is the sum of the fields scattered at the jth particle. Subscripts l and j imply the transition from the l to the j coordinate system.

The incident field is defined as

$$\mathbf{E_0}(j) = -\sum_{n=1}^{\infty} \sum_{m=-n}^{n} i E_{mn} [p_{mn}^{jo,j} \mathbf{N}_{mn}^1(kr) + q_{mn}^{jo,j} \mathbf{M}_{mn}^1(kr)]. \tag{6.70}$$

Waves are incident relative to the center of each jth particle (i.e., in the jth system of coordinates). The expansion coefficients of the incident plane electromagnetic wave have the form [12]:

$$p_{mn}^{jo,j} = 4\pi(-1)^m i^n d_n \mathbf{C}_{mn}^*(\theta_{inc}) \mathbf{E}_{inc}(\mathbf{k}_{inc}, \mathbf{r}_{j0,j}) \exp(-im\varphi_{inc}),$$

$$q_{mn}^{jo,j} = 4\pi(-1)^m i^{n-1} d_n \mathbf{B}_{mn}^*(\theta_{inc}) \mathbf{E}_{inc}(\mathbf{k}_{inc}, \mathbf{r}_{j0,j}) \exp(-im\varphi_{inc}),$$

where $\mathbf{E}_{inc}(\mathbf{k}_{inc}, \mathbf{r}_{j0,j})$ is the linear polarization vector, $\mathbf{k}_{inc}$ is the wave vector, the asterisk indicates complex conjugation, d_n, $\mathbf{B}_{mn}$ and $\mathbf{C}_{mn}$ are defined by formulas (6.7)–(6.9). Let us write the expression for the scattered field:

$$\mathbf{E_s}(l, j) = -\sum_{n=1}^{\infty} \sum_{m=-n}^{n} i E_{mn} [p_{mn}^{l,j} \mathbf{N}_{mn}^1 + q_{mn}^{l,j} \mathbf{M}_{mn}^1], \tag{6.71}$$

where coefficients $p_{mn}^{l,j}$, $q_{mn}^{l,j}$ are defined in Chap. 3.

Combining expressions (6.35), (6.69) and (6.70) and taking into account relation (6.38), we obtain an infinite system of linear algebraic equations for the jth particle of an arbitrary shape:

$$\begin{pmatrix} a^j \\ b^j \end{pmatrix} = T_2^j \left[\begin{pmatrix} p^{jo,j} \\ q^{jo,j} \end{pmatrix} + \sum_{l \neq j} \begin{pmatrix} A(l,j) & B(l,j) \\ B(l,j) & A(l,j) \end{pmatrix} \begin{pmatrix} a^j \\ b^j \end{pmatrix} \right], \quad (6.72)$$

where coefficients $A(l,j)$, $B(l,j)$ are defined in Chap. 3. The solution of the system of linear equations (6.72) was carried out using the stable algorithm of biconjugate gradients (BiCGSTAB). Having determined coefficients a_{mn}^j, b_{mn}^j and from this system, we can write the expression for the scattered field in the main system of coordinates

$$\mathbf{E}_s = \sum_{n=1}^{\infty} \sum_{m=-n}^{n} i \, E_{mn} [a_{mn} \mathbf{N}_{mn}^3 + b_{mn} \mathbf{M}_{mn}^3]. \quad (6.73)$$

The component-wise form of the scattered field is given by

$$E_{s\theta} \sim E_0 \frac{e^{ikr}}{-ikr} \sum_{n=1}^{\infty} \sum_{m=-n}^{n} (2n+1) \frac{(n-m)!}{(n+m)!} [a_{mn} \tau_{mn} + b_{mn} \pi_{mn}] e^{im\phi}, \quad (6.74)$$

$$E_{s\phi} \sim E_0 \frac{e^{ikr}}{-ikr} \sum_{n=1}^{\infty} \sum_{m=-n}^{n} (2n+1) \frac{(n-m)!}{(n+m)!} [a_{mn} \pi_{mn} + b_{mn} \tau_{mn}] e^{im\phi}, \quad (6.75)$$

where

$$\tau_{mn} = \frac{\partial}{\partial \theta} P_n^m(\cos \theta), \ \pi_{mn} = \frac{m}{\sin \theta} P_n^m(\cos \theta).$$

Symbol ($\sim$) indicates that expressions (6.74) and (6.75) following from (6.73) are treated asymptotically for ($kr \gg 1$). Since we consider here the scattering at large distance from the jth particle, the electric vectors of the scattered field are parallel to the electric vector of the incident field; i.e., only the θ component differs from zero in the far zone, and expressions (6.74) and (6.75) can be simplified:

$$E_{s\theta} \sim E_0 \frac{e^{ikr}}{-ikr} \sum_{n=1}^{\infty} \sum_{m=-n}^{n} \frac{(2n+1)}{n(n+1)} [a_{mn} \tau_n + b_{mn} \pi_n] \quad (6.76)$$

$$E_{s\phi} \sim E_0 \frac{e^{ikr}}{-ikr} \sum_{n=1}^{\infty} \sum_{m=-n}^{n} \frac{(2n+1)}{n(n+1)} [a_{mn} \pi_n + b_{mn} \tau_n], \quad (6.77)$$

where

$$\tau_n = \frac{\partial}{\partial \theta} P_n(\cos \theta), \ \pi_n = \frac{1}{\sin \theta} P_n(\cos \theta)$$

Analogous expressions can also be obtained for magnetic field components $H_{s\phi}$ and $H_{s\theta}$.

6.5 Reflection of a Plane Wave from a Layer with a Slowly Varying Thickness

Let us consider the following optical scheme. The system consists of four regions with difference refractive indices(epidermis, the upper layer of the dermis, blood cells,the lower layer of the dermis) (see Fig. 6.2).

To attain the best agreement between the structure and the actual object under investigation, we represent the interfaces between the layers of the model medium in the form of certain surfaces $z_i = H_i(x, y), i = \overline{1, 3}$.

Let us suppose that a plane s- or p- polarized wave is incident on the layer at an angle θ. We consider only the case of the p polarization. We must find the reflected field.

We will seek the reflected field in the form of waves with slowly varying amplitudes and rapidly oscillating phases

$$E_1 = \exp\left(\frac{i}{\varepsilon}\tau_{inc}(\xi_1, \xi_2, \xi_3)\right) + \exp\left(\frac{i}{\varepsilon}\tau_{1ref}(\xi_1, \xi_2, \xi_3)\right) A(\xi_1, \xi_2, \xi_3, \varepsilon_x, \varepsilon_y),$$
(6.78)

$$E_2 = \exp\left(\frac{i}{\varepsilon}\tau_{2tr}(\xi_1, \xi_2, \xi_3)\right) B^+(\xi_1, \xi_2, \xi_3, \varepsilon_x, \varepsilon_y)+$$

$$+ \exp\left(\frac{i}{\varepsilon}\tau_{3ref}(\xi_1, \xi_2, \xi_3)\right) B^-(\xi_1, \xi_2, \xi_3, \varepsilon_x, \varepsilon_y),$$
(6.79)

Fig. 6.2 Schematic diagram of biological medium

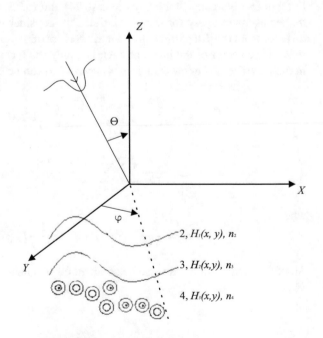

$$E_3 = \exp\left(\frac{i}{\varepsilon}\tau_{3elap}(\xi_1, \xi_2, \xi_3)\right) C^+(\xi_1, \xi_2, \xi_3, \varepsilon_x, \varepsilon_y) +$$

$$+ \exp\left(\frac{i}{\varepsilon}\tau_{4ref}(\xi_1, \xi_2, \xi_3)\right) C^-(\xi_1, \xi_2, \xi_3, \varepsilon_x, \varepsilon_y), \tag{6.80}$$

$$E_4 = \exp\left(\frac{i}{\varepsilon}\tau_{4elap}(\xi_1, \xi_2, \xi_3)\right) D^+(\xi_1, \xi_2, \xi_3, \varepsilon_x, \varepsilon_y) +$$

$$\exp\left(\frac{i}{\varepsilon}\tau_{5ref}(\xi_1, \xi_2, \xi_3)\right) D^-(\xi_1, \xi_2, \xi_3, \varepsilon_x, \varepsilon_y) +$$

$$+ E_{4\theta scat}(\xi_1, \xi_2, \xi_3) \tag{6.81}$$

$$E_5 = \exp\left(\frac{i}{\varepsilon}\tau_{5elap}(\xi_1, \xi_2, \xi_3)\right) E(\xi_1, \xi_2, \xi_3, \varepsilon_x, \varepsilon_y), \tag{6.82}$$

where $E_{4\theta scat}$ is defined by formula (6.76) and $\tau_{1inc}, \tau_{1ref}, \tau_{2elap}, \tau_{3ref}, \tau_{3elap}, \tau_{4ref}, \tau_{4elap}$ are defined in Chap. 4. Amplitudes A, $B^\pm$, $C^\pm$, $D^\pm$, E are sought in the form of power series in small parameter ε_x, ε_y, the expressions for the amplitudes are defined analogously to the method described in Chap. 4.

Substitution of expressions (8.1)–(8.5) into (4.6)–(4.11) generates a recurrence system of equations. For the reflected field, this system leads to reflection coefficient A. The expression for the reflection of a Gaussian beam with an arbitrary cross section is defined analogously to the method described in Chap. 4.

6.6 Spectrum of Action of Laser Radiation on the Hemoglobin Derivatives

Let us consider the mathematical simulation of the spectral efficiency of light absorption by the main blood hemoglobin derivatives: oxyhemoglobin(HbO_2) and deoxyhemoglobin(Hb) of human blood in the upper layers of the human dermis.

It should be noted that the mechanism of action of laser radiation on biological structures are not completely clear as yet; several processes (namely, photoinduced dissociation of oxyhemoglobin of blood, which is accompanied by the molecular oxygen liberation and a local increase in its concentration in blood [20, 21]); as a result of this photochemical reaction, deoxyhemoglobin is formed, and an opto-oxygen effect is observed [20–22], which is responsible for the liberation of singlet oxygen from triplet oxygen dissolved in the cells. It should be noted that the above processes depend on the efficiency of light absorption by blood and, hence, on the radiation wavelength and the radiation power density at a given depth.

In analysis of the efficiency of photochemical and photophysical processes, we will use the concept of action spectrum. The spectrum of light action on a tissue component is the total power of radiation absorbed by this component in unit volume of the medium when monochromatic light of unit power density is incident on the surface of the medium [20]:

$$K_{HbO_2}(\lambda) = C_v \cdot H \cdot f \cdot S \cdot \mu_{a(HbO_2)}(\lambda) \times$$

$$\times \int_{4\pi} I(\lambda, x, y, m_\tau^j, x_\tau^j, \Omega) d\Omega, \qquad (6.83)$$

$$K_{Hb}(\lambda) = C_v \cdot H \cdot f \cdot (1 - S) \cdot \mu_{a(Hb)}(\lambda) \times$$

$$\times \int_{4\pi} I(\lambda, x, y, m_\tau^j, x_\tau^j, \Omega) d\Omega, \qquad (6.84)$$

$$K_{blood}(\lambda) = K_{HbO_2}(\lambda) + K_{Hb}(\lambda), \qquad (6.85)$$

where $d\Omega = \sin\theta d\theta d\varphi$ is the solid angle, K_{HbO_2}, K_{Hb}, K_{blood} are the action spectra of light on oxyhemoglobin, deoxyhemoglobin, and blood, respectively, H is the capillary hematocrit(volume concentration of erythrocytes in blood); f is the volume concentration of hemoglobin in erythrocytes, S is the degree of oxygenation of blood (ratio of the concentration of oxyhemoglobin to the total concentration of hemoglobin), $I(x, y, m_\tau^j, x_\tau^j, \Omega)$ is the intensity and defined by formula (4.69), $\mu_{a(HbO_2)}(\lambda)$ are the absorption spectra of oxyhemoglobin, $\mu_{a(Hb)}(\lambda)$ are the absorption spectra of deoxyhemoglobin [23], $m_\tau^j = N_\tau^j/n_o$, N_τ^j is the complex refractive index of the jth particle for the τ concentric layer, n_o is the refractive index of the surrounding medium, $x_\tau^j = ka_\tau^j$, $j = \overline{1...N}$, $\tau = \overline{1, 2}$, where a_τ^j-is the radius of the jth particle with concentric layer τ.

Thus, at this stage, we use formulas (6.83)–(6.85) to connect the action spectra of oxyhemoglobin (HbO₂) and deoxyhemoglobin (Hb) and blood of the biotissue under investigation as functions of the wavelength of laser radiation taking into account the electrophysical parameters of the biological structure being simulated such as the real and imaginary parts of the refractive indices and sizes.

Let us consider the choice of the values for hematocrit. It was shown in [24] that the hematocrit in capillaries can be smaller than in arteries and veins; for example, when blood flows into capillaries from the artery through a narrow arteriole, the hematocrit can decrease from 0.5 to 0.068. Such a decrease in the hematocrit is known as the Fahraeus effect [25]. Such a variation of the hematocrit can apparently be explained by the following circumstances [20]:

1. A considerable portion of blood flows from the artery to a microvessel from the near-wall region in which the plasma concentration is elevated. Note that the specific manifestations of the Fahraeus effect depend on various characteristics of the blood flow and the metabolic activity of tissues [20]. It was shown in [24] that when the blood flows through the expanded arteriole, the hematocrit in the capillary decreases

from 0.5 to 0.38. However, in [26] the inverse Fahraeus effect was observed, when the values of the hematocrit in a capillary were higher than in great vessels.

2. Insufficient deformation of erythrocytes prevents their flow into a narrow capillary.

Thus, according to [20, 26], the value of the hematocrit can be chosen as 0.4.

6.7 Numerical Calculations for a Model Medium and Conclusions

Let us consider a model medium with the following characteristics. Typical layer thicknesses are equal to $d_2 = 65 \cdot 10^{-6}$, $d_3 = 565 \cdot 10^{-6}$, $d_4 = 90 \cdot 10^{-6}$, $n_1^\circ = 1$, $\chi_1 = 0$, $\chi_2 = \chi_3 = \chi_4 = \chi_5 = 10^{-5}$, refractive indices of the layers are $n_2^\circ = 1.50$, $n_3^\circ = 1.40$, $n_4^\circ = 1.35$, $n_5^\circ = 1.40$ and the following values of parameters are $a_1 = -0.0024$, $b_1 = 0.020$, $a_2 = 0.021$, $b_2 = 0.030$, $a_3 = 0.041$, $b_3 = 0.051$, $c_1 = c_2 = c_3 = 10^{-2}$. The values of parameters for the interfaces between the layers are chosen so that the shape of the surface is maximally close to the shape of the boundary of the corresponding layer in the structure of the normal human dermis, and wavelength is $\lambda = 0.63\,\mu m$ (center of the line of a He−Ne laser).

Since the erythrocyte contains no cell organelles, its cellular membrane is very thin and does not noticeably affect the scattering of light; consequently, the erythrocyte can be treated as a homogeneous scatterer. Thus, our computations were performed for monolayer spherulated particles simulating erythrocytes; the number of particles in the layer being simulated was assumed to be ten for the following parameters: the relative refractive index for the first five spherulated erythrocytes was assumed to be $1.035 + 10^{-5}i$; for the remaining erythrocytes, it was set as $1.033 + 10^{-5}i$, for a particle radius of $4.3\,\mu m$, $H = 0.4$, $f = 0.08$, $S = 0.75$, $C_v = 0.0595$ [27]. All computations were performed up to 32 decimal places.

Figure 6.3a, b illustrates the distribution of radiation intensity for multilayer medium absorbing and scattering light, which simulates human dermis for specific electrophysical and geometrical characteristics of the biological structure being simulated. The dependences of the laser radiation intensity on the refractive index and absorption coefficient of the epidermis for various electrophysical parameters of the biotissue under investigation are shown in Fig. 6.4a, b.

It should be noted that the model constructed here is quite sensitive to variations of the refractive index of the biological structure being simulated; the model also permits the variation of electrophysical parameters of the biological sample under investigation, its geometrical parameters, and the establishment of the relation between these parameters and the biological properties of the biotissue being simulated. Thus, this model can be used for measuring in vivo the spectral differences between the normal and pathological tissues for determining pathological changes in the biosamples under investigation, which are associated with a variation of electrophysical properties of epidermis and blood corpuscles in the upper layer of the dermis.

Fig. 6.3 Radiation intensity distribution for a multilayer medium absorbing and scattering light for specific parameters $\theta = 0°$, $\varphi = 0°$, $\psi = 0°$ and **a** refractive index of epidermis is $1.303 + 10^{-3}i$ and the refractive index of the dermis is $1.301 + 10^{-3}i$; **b** refractive index of epidermis is $1.303 + 10^{-4}i$ and the refractive index of the dermis is $1.301 + 10^{-4}i$

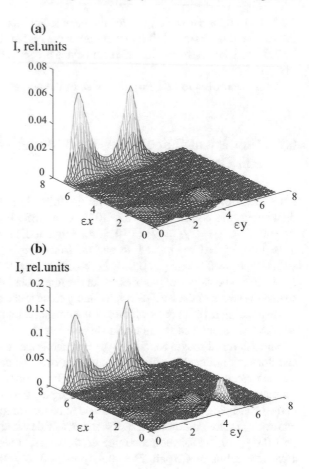

Figure 6.5a, b shows the normalized spectra of action of laser radiation on oxy- and deoxyhemoglobin. The simulation of the action spectra of the laser radiation power absorbed by oxyhemoglobin and deoxihemoglobin of blood is performed using the theory of the T-matrix in the spectral range from 300 to 800 nm. The choice of this spectral interval is dictated by the fact that in most available methods, the transparency window in the wavelength range from 650 to 1200 nm is used for optical probing of biotissues [28]. Note that the spectral interval from 400 to 600 nm is diagnostic because the main absorption bands of blood (Sore band 420 nm, absorption bands α and β of oxyhemoglobin at 545 and 575 nm) lie in this spectral interval.

The spectral dependences of the refractive indices of the dermis and epidermis were described by the following expressions [29]:

$$n(\lambda) = 1.30904 - \frac{434.60068}{\lambda^2} + \frac{1.60647 \cdot 10^9}{\lambda^4} - \frac{1.28111 \cdot 10^{14}}{\lambda^6}$$

Fig. 6.4 Dependences of the laser radiation intensity on the refractive index and absorption coefficient of the epidermis. The refractive index of the upper layer of the dermis is $1.6 + 0.001i$ **(a)** and $1.36 + 0.0001i$ **(b)**

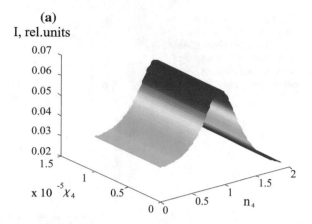

(a)
I, rel.units

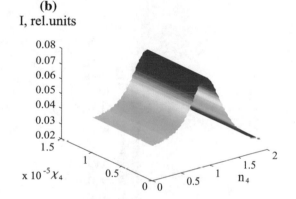

(b)
I, rel.units

$$n(\lambda) = 1.68395 - \frac{1.87232 \cdot 10^4}{\lambda^2} + \frac{1.09644 \cdot 10^{10}}{\lambda^4} - \frac{8.64842 \cdot 10^{14}}{\lambda^6}$$

Note that analogous results of analysis of the action spectra of laser radiation on oxy- and deoxihemoglobin were obtained in [20]. Certain differences between the results given in [20] and in Fig. 6.5a, b are due, first, to the use of oxy- and deoxihemoglobin for the initial absorption spectra; second, the knowledge of the spectral dependences of the refractive index of the epidermis, dermis, and the average refractive index of blood corpuscles is required for a more adequate description of propagation of laser radiation in biological media, while in our calculations, the averaged refractive index of the epidermis, dermis, and the averaged refractive index of blood corpuscles were used.

Thus, the model constructed in this study makes it possible not only to select optimal wavelengths for effective action of laser radiation on biological structures, but also to analyze the effectiveness of absorption not only by blood, but also by biotissues like melanin of the epidermis. The above dependences can be used for predicting the changes in the optical properties of blood in the capillary channel, which are

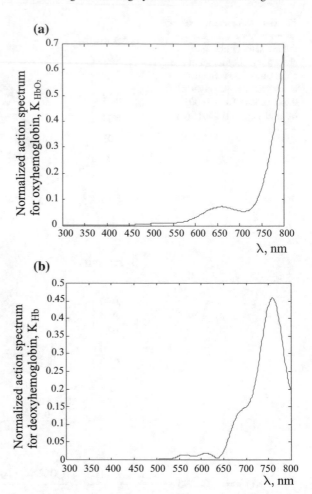

Fig. 6.5 **a** Normalized action spectra for laser radiation for oxyhemoglobin; **b** normalized action spectra for laser radiation for deoxihemoglobin

associated with various biophysical, biochemical, and physiological processes, and can be computed for lasers with other parameters; the quantitative estimates obtained in this study can be applied for processing and interpreting experimental data.

We have described the mathematical model for calculating the optical characteristics and for analyzing the biophysical processes of propagation of light in a multilayer biotissue in the case of the interaction with noncoagulating laser radiation. The model was implemented in the form of a software package, which makes it possible to vary automatically the composition of biological objects, their electrophysical parameters, characteristic thicknesses of layers, as well as characteristic sizes of various biological structures under investigation on the same setup for recording the dependence between these parameters. This makes the software developed here an effective and convenient tool for investigations in biomedical optics.

References

1. E.Y. Eremina, Y.A. Eremin, T. Wriedt, Analysis of light scattering by erythrocyte based on discrete sources method. Opt. Commun. **244**, 15–23 (2005)
2. E.Y. Eremina, Y.A. Eremin, T. Wriedt, Different shape models for erythrocyte: light scattering analysis based on the discrete sources method. J. Quant. Spectrosc. Radiat. Transf. **102**, 3–10 (2006)
3. E.Y. Eremina, T. Wriedt, Light scattering analysis by a particle of extreme shape via discrete sources method. J. Quant. Spectrosc. Radiat. Transf. **89**, 67–77 (2004)
4. M.I. Mishchenko, W.J. Wiscombe, L.D. Travis, *Light Scattering by No Spherical Particles: Theory, Measurements and Applications* (Academic press, San-Diego, 2000), Ch 2, pp. 29–60
5. P.C. Waterman, Matrix formulation of electromagnetic scattering. Proc. IEEE. **53**(8), 805–812 (1969)
6. P.C. Waterman, Symmetry, unitarity and geometry in electromagnetic scattering. Phys. Rev. D **3**(4), 825–839 (1971)
7. K.G. Kulikov, Light scattering by dielectric bodies of arbitrary shape with the application to biophysical problem. in *SPIE Photonics Europe, 16–19 April 2012, Belgium, Brussels, 2012*
8. K.G. Kulikov, Light scattering by dielectric bodies of irregular shape in a layered medium in problems of biomedical optics: I. Theory Comput. Model Tech. Phys. **57**(12), 1623–1631 (2012)
9. K.G. Kulikov, Light scattering by dielectric bodies of irregular shape in a layered medium in problems of biomedical optics: II. Numer. Anal. Tech. Phys. **8**(12), 24–28 (2012)
10. J.M. Steinke, A.P. Shepherd, Comparison of Mie theory and the light scattering of red blood cells. Appl. Opt. **27**, 4027–4033 (1988)
11. A.N. Yaroslavsky, T. Goldbach, H. Schwarzmaier, Influence of the scattering phase function approximation on the optical properties of blood determined from the integrating sphere measurements. J. Biomed. Opt. **4**(1), 47–53 (1999)
12. L. Tsang, J.A. Rony, R.T. Shin, in *Theory of Microwave Remote Sensing* (New York, 1985)
13. G. Korn, T. Korn, in *Handbook of Mathematics for Scientists and Engineers* (Moscow, 1973)
14. A. Doicu, T. Wriedt, Y.A. Eremin, *Light Scattering by Systems of Particles, Null-Field Method with Discrete Sources: Theory and Programs* (Springer, Berlin, New York, 2006)
15. D.S. Wang, P.W. Barber, Scattering by inhomogeneous nonspherical objects. Appl. Opt. **18**, 1190–1198 (1979)
16. I.M Gel'fand, R.A. Minlos, Z.Ya. Shapiro, in *Representations of the Rotation Group and Lorentz Group and Their Applications* (Moscow, 1958)
17. D.A. Varshalovich, A.N. Moskaliev, V.K. Kherson, in *Quantum Theory of Angular Momentum* (Leningrad, 1975)
18. M.I. Mishchenko, L.D. Travis, T-matrix computations of light scattering by large spheroidal particles. Opt. Commun. **109**, 16–21 (1994)
19. M.I. Mishchenko, L.D. Travis, Capabilities and limitations of a current FORTRAN implementation of the T-matrix method for randomly oriented, rotationally symmetric scatterers. J. Quant. Spectrosc. Radiat. Transfer. **60**, 309–324 (1998)
20. V.V. Barun, A.P. Ivanov, The light absorption of blood during low-intensity laser irradiation of the skin. Quantum Electron. **40**(4), 371–378 (2010)
21. M.M. Asimov, R.M. Asimov, A.N. Rubinov, Action spectrum of the laser radiation on hemoglobin of blood vessels in the skin. J. Appl. Spectrosc. **65**, 919 (1998)
22. S.D. Zakharov, A.V. Ivanov, Optical oxygen effect in cells and prospects of using in the therapy of tumors. Quantum Electron. **29**, 192 (1999)
23. http://omlc.ogi.edu/spectra/hemoglobin/index.html
24. B.R. Duling, C. Desjardins, Capillary haematocrit - what does it mean. News Physiol. Sci. **2**, 66 (1987)
25. R. Fahraeus, The suspension stability of blood. Physiol. Rev. **9**, 241 (1929)
26. R.T. Yen, Y.C. Fung, Inversion of Fahraeus effect and effect of mainstream flow on capillary hematocrit. J. Appl. Physiol. **42**, 578 (1977)

27. I.V. Meglinski, Simulation of the reflectance spectra of optical radiation from a randomly inhomogeneous multilayer strongly scattering and absorbing light environments using the Monte Carlo. Quantum Electron. **31**(12), 1101–1107 (2001)
28. V.V. Tuchin, *Lasers and Fiber Optics in Biomedical Studies* (Saratovsky Univ, Saratov, 1998)
29. A.N. Bashkatov, Controlling the optical properties of biological tissues upon exposure to osmotically active immersion liquids. Ph.D. dissertation, Saratov, 2002

Chapter 7
Modeling of the Optical Characteristics Fibrillar Structure

Abstract We describe the mathematical model, which allows us to vary the electrical parameters and structure of the simulated biological tissue with fibrillar structure for case in vivo.

7.1 Introduction

At present optical diagnostic methods tissues occupy a leading position because of their high information content, and also their relative simplicity and low cost.

There are numerous diagnostic techniques, such as optical coherence tomography, confocal microscopy, fluorescence spectroscopy, diffuse optical tomography, that require knowledge of the optical properties and the dynamics of diffusion of various of medicinal substances in various biological tissues.

In spite of significant advances in the development of fundamental bases and practical applications of optical methods tissues, actual problems at present are the increasing effects and expanded functionality possibilities of existing diagnostic techniques.

Note that at present the degree of development of representations about the propagation of light in multiple scattering media with a fibrillar structure that consist of partially oriented fibers are insufficient.

Such objects represent the considerable interest for biomedical applications. We note some articles devoted to research optical anisotropy of the tissue with fibrillar structure.

In [1] are presented the results of the theoretical analysis optical anisotropy of multiply scattering fibrillar tissues, conducted on the basis of models of effective anisotropic medium with experimental data on double refraction in vivo derma of rat. The article [2] is devoted to the question of dynamics immersion blooming different types of biological fabrics, construction models and methods to describe the propagation light emission with different types of polarization through anisotropic tissue. In the article [3] one studies the problem of anisotropic scattering of light in biological tissues, which have cylindrical structure (e.g., collagen) by the Monte-Carlo method.

© Springer International Publishing AG, part of Springer Nature 2018 131
K. Kulikov and T. Koshlan, *Laser Interaction with Heterogeneous Biological Tissue*, Biological and Medical Physics, Biomedical Engineering,
https://doi.org/10.1007/978-3-319-94114-1_7

The obtained results make it possible to determine the optical properties of the tissue, they are also useful for the diagnosis of early changes tissues.

Thus becomes important to use the mathematical modeling of physical processes proceeding in biological samples of different types in conditions of laser irradiation.

The problem consists of several parts. In the first part, we consider the problem of light scattering on a system of parallel dielectric cylinders, modeling collagen fibers. In the second part we consider the problem of reflection of a Gaussian beam with an arbitrary cross-section plane wave from a smoothly irregular layer modeling biological tissue with fibrillar structure. In the third part of the numerical simulations we investigate the question of electrical characteristics of the biological sample.

7.2   Scattering on a Parallel Cylinders

In this section, we consider the distribution of polarized radiation in multiple scattering media the example of the dermis.

The structure of the dermis are collagen fibers, consisting of parallel beams of an average thickness 50–100 nm, connected by glycosaminoglycans and proteoglycans [4], then applied to the analysis of the effect of the morphological characteristics of multiple scattering in randomly inhomogeneous media with fibrillar structure regarded in the modeled medium that consists of parallel dielectric cylinders with identical refraction complex coefficient are n_{cyl} and radius a.

In this case, the cylinders are distributed in an isotropic dielectric medium with a complex refractive refraction are n_o. The distance between the cylinder and the wavelength incident radiation are comparable and we believe that the cylinders are oriented along the Z-axis.

Let the cylinders at an angle θ be incident by the plane polarized wave in this case, the Cartesian coordinate system $OXZY$ is used as a fixed reference system, φ is the azimuthal angle. We consider only the simple harmonic time dependence of the angular frequency ω, and the factor $\exp(-i\omega t)$ is omitted.

Note that cylindrical coordinates of primary field incident wave are $\mathbf{E}$, $\mathbf{H}$ admits a representation as superposition of electric and magnetic fields types.

Thus in solving the problem, we consider two cases of the polarization of the incident wave. Note that the scattered on the cylinders field can be found, depending on the polarization primary field through the magnetic U and the electric V Hertz potential. In this case, the values potentials U, V associated with the vector $\mathbf{E}$, $\mathbf{H}$ by the following relations:

$$\mathbf{E} = \frac{i}{k} \nabla \times \nabla \times (\mathbf{e_z}U) + \nabla \times (\mathbf{e_z}V), \tag{7.1}$$

$$\mathbf{H} = -n_o \nabla \times (\mathbf{e_z}U) + \frac{i}{k} \nabla \times \nabla \times (\mathbf{e_z}V) \tag{7.2}$$

and satisfy the wave equations:

$$\Delta U + n_o^2 k^2 U = 0,$$

$$\Delta V + n_o^2 k^2 V = 0.$$

We write the relations for longitudinal, azimuthal and radial component of the electric and magnetic fields which expressed through U, V in cylindrical coordinates (ρ, φ, z):

$$E_\rho = \frac{\partial^2 U}{\partial \rho \partial z}, \quad E_\varphi = \frac{1}{\rho} \frac{\partial^2 U}{\partial \varphi \partial z}, \quad E_z = \frac{\partial^2 U}{\partial z^2} + k^2 n_o^2 U, \tag{7.3}$$

$$H_\rho = -\frac{n_o}{\rho} \frac{\partial U}{\partial \varphi}, \quad H_\varphi = n_o \frac{\partial U}{\partial \rho}, \quad H_z = 0 \tag{7.4}$$

$$H_\rho = \frac{\partial^2 V}{\partial \rho \partial z}, \quad H_\varphi = \frac{1}{\rho} \frac{\partial^2 V}{\partial \varphi \partial z}, \quad H_z = \frac{\partial^2 V}{\partial z^2} + k^2 n_o^2 V, \tag{7.5}$$

$$E_\rho = -\frac{n_o}{\rho} \frac{\partial V}{\partial \varphi}, \quad E_\varphi = n_o \frac{\partial V}{\partial \rho}, \quad E_z = 0, \tag{7.6}$$

where U, V is

$$\frac{1}{\rho} \left[\frac{\partial}{\partial \rho} \left[\rho \frac{\partial U}{\partial \rho} \right] + \frac{1}{\rho} \left[\frac{\partial^2 U}{\partial \varphi^2} \right] \right] + \frac{\partial^2 U}{\partial z^2} + k^2 n_o^2 U = 0,$$

$$\frac{1}{\rho} \left[\frac{\partial}{\partial \rho} \left[\rho \frac{\partial V}{\partial \rho} \right] + \frac{1}{\rho} \left[\frac{\partial^2 V}{\partial \varphi^2} \right] \right] + \frac{\partial^2 V}{\partial z^2} + k^2 n_o^2 V = 0.$$

The field incident on the surface relative to jth cylinder consists of several parts: the initial incident waves, the primary wave scattered by the jth cylinder and wave scattered on all other cylinders. Then we can write general expression for the scalar field potential functions U, V:

$$U^j = U^j_{inc}(\mathbf{R}_{jp}) + U^j_{scat}(\mathbf{R}_{jp}) + \sum_{k \neq j}^{N} U^k_{scat}(\mathbf{R}_{kp}), \tag{7.7}$$

$$V^j = V^j_{inc}(\mathbf{R}_{jp}) + V^j_{scat}(\mathbf{R}_{jp}) + \sum_{k \neq j}^{N} V^k_{scat}(\mathbf{R}_{kp}), \tag{7.8}$$

where $U^j_{inc}(\mathbf{R}_{jp}), V^j_{inc}(\mathbf{R}_{jp})$ are potential incident field on the surface of the jth cylinder, $U^j_{scat}(\mathbf{R}_{jp})$, $V^j_{scat}(\mathbf{R}_{jp})$ are potential of primary scattered wave at j-th cylinder, $U^k_{scat}(\mathbf{R}_{kp})$, $V^k_{scat}(\mathbf{R}_{kp})$ are potential of the scattered field at all other cylinders.

Expressions of the form (7.7)–(7.8) can be written as

$$\begin{pmatrix} U^j \\ V^j \end{pmatrix} = \begin{pmatrix} U^j_{inc}(\mathbf{R}_{jp}) \\ V^j_{inc}(\mathbf{R}_{jp}) \end{pmatrix} + \begin{pmatrix} U^j_{scat}(\mathbf{R}_{jp}) \\ V^j_{scat}(\mathbf{R}_{jp}) \end{pmatrix} + \sum_{k \neq j}^{N} \begin{pmatrix} U^k_{scat}(\mathbf{R}_{kp}) \\ V^k_{scat}(\mathbf{R}_{kp}) \end{pmatrix}. \tag{7.9}$$

The potential of the incident field for the case p- and s- polarization is the following form

$$U^j_{inc}(\mathbf{R}_{jp}) = e^{-ikz \sin \theta} e^{-ik\mathbf{R}_{jp}}, \ V^j_{inc}(\mathbf{R}_{jp}) = 0, \tag{7.10}$$

$$V^j_{inc}(\mathbf{R}_{jp}) = e^{-ikz \sin \theta} e^{-ik\mathbf{R}_{jp}}, \ U^j_{inc}(\mathbf{R}_{jp}) = 0. \tag{7.11}$$

Using the expansion of a plane wave by a cylindrical wave functions and substituting (7.12) for (7.10)–(7.11), we obtain

$$e^{ik\rho \sin \theta} = \sum_{n=\infty}^{\infty} (-i)^n J_n(k\rho) e^{in\theta}, \tag{7.12}$$

$$U^j_{inc}(\mathbf{R}_{jp}) = e^{-ikz \sin \theta} \sum_{n=\infty}^{\infty} (-i)^n J_n(k\mathbf{R}_{jp} \cos \theta n_o) e^{in\varphi} e^{in\gamma_{jp}}, \ V^j_{inc}(\mathbf{R}_{jp}) = 0, \tag{7.13}$$

$$V^j_{inc}(\mathbf{R}_{jp}) = e^{-ikz \sin \theta} \sum_{n=\infty}^{\infty} (-i)^n J_n(k\mathbf{R}_{jp} \cos \theta n_o) e^{in\varphi} e^{in\gamma_{jp}}, \ U^j_{inc}(\mathbf{R}_{jp}) = 0, \tag{7.14}$$

γ_{jp} is angle to jth cylinder (see Fig. 7.1).

Combining the expressions (7.13)–(7.14) we obtain

Fig. 7.1 Geometric illustration of scattering by cylinders

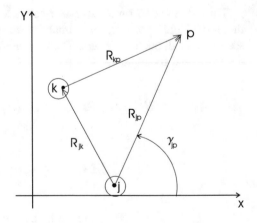

$$\begin{pmatrix} U_{inc}^j(\mathbf{R}_{jp}) \\ V_{inc}^j(\mathbf{R}_{jp}) \end{pmatrix} = \begin{pmatrix} \alpha \\ 1-\alpha \end{pmatrix} e^{-ikz\sin\theta} \sum_{n=\infty}^{\infty}(-i)^n J_n(k\mathbf{R}_{jp}\cos\theta n_o)e^{in\varphi}e^{in\gamma_{jp}}, \quad (7.15)$$

where $\alpha = 1$ for the case p polarization, and $\alpha = 0$ for the case s polarization. Analogous to [5] we write the expression for the scattered field at the jth cylinder through scalar potential functions:

$$U_{scat}^j(\mathbf{R}_{jp}) = -e^{-ikz\sin\theta}\sum_{n=\infty}^{\infty}(-i)^n H_n^2(k\mathbf{R}_{jp}\cos\theta n_o)e^{in\gamma_{jp}}a_n^j, \quad (7.16)$$

$$V_{scat}^j(\mathbf{R}_{jp}) = -e^{-ikz\sin\theta}\sum_{n=\infty}^{\infty}(-i)^n H_n^2(k\mathbf{R}_{jp}\cos\theta n_o)e^{in\gamma_{jp}}b_n^j, \quad (7.17)$$

or

$$\begin{pmatrix} U_{scat}^j(\mathbf{R}_{jp}) \\ V_{scat}^j(\mathbf{R}_{jp}) \end{pmatrix} = -e^{-ikz\sin\theta}\sum_{n=\infty}^{\infty}(-i)^n H_n^2(k\mathbf{R}_{jp}\cos\theta n_o)e^{in\gamma_{jp}}\begin{pmatrix} a_n^j \\ b_n^j \end{pmatrix}, \quad (7.18)$$

where

$$\begin{pmatrix} a_n^j \\ b_n^j \end{pmatrix} = \begin{pmatrix} \alpha a_n^{j\,I} + (1-\alpha)a_n^{j\,II} \\ \alpha b_n^{j\,I} + (1-\alpha)b_n^{j\,II} \end{pmatrix}, \quad (7.19)$$

$a_n^{j\,I}, b_n^{j\,I}\ a_n^{j\,II}, \ a_n^{j\,I}, b_n^{j\,I}\ a_n^{j\,II}, \ b_n^{j\,II}$ are scattering coefficients on the cylinder for p- polarization and for s- polarization.

By the addition theorem for a pair of cylinders we have [6]

$$e^{in\psi_k} H_n^2(k\mathbf{R}_{kp}\cos\theta n_o) = \sum_{s=-\infty}^{\infty}\sum_{n=-\infty}^{\infty} H_{s-n}^2(k\mathbf{R}_{jk}\cos\theta n_o))\times$$

$$\times J_n(k\mathbf{R}_{jp}\cos\theta n_o))e^{is\psi_s}. \quad (7.20)$$

The potential of the scattered field for all other cylinders considering expression (7.20) is

$$U_{scat}^k(\mathbf{R}_{kp}) = -e^{-ikz\sin\theta}\sum_{k\neq j}^{N}\sum_{s=-\infty}^{\infty}\sum_{n=-\infty}^{\infty}(-i)^s e^{in\gamma_{jp}}e^{in(s-n)\gamma_{kj}}H_{s-n}^2(k\mathbf{R}_{jk}\cos\theta n_o)\times$$

$$\times J_n(k\mathbf{R}_{jp}\cos\theta n_o)a_s^k \quad (7.21)$$

$$V_{scat}^k(\mathbf{R}_{kp}) = -e^{-ikz\sin\theta} \sum_{k\neq j}^{N} \sum_{s=-\infty}^{\infty} \sum_{n=-\infty}^{\infty} (-i)^s e^{in\gamma_{jp}} e^{in(s-n)\gamma_{kj}} H_{s-n}^2(k\mathbf{R}_{jk}\cos\theta n_o) \times$$

$$\times J_n(k\mathbf{R}_{jp}\cos\theta n_o)b_s^k \tag{7.22}$$

or

$$\begin{pmatrix} U_{scat}^j(\mathbf{R}_{kp}) \\ V_{scat}^j(\mathbf{R}_{kp}) \end{pmatrix} = -e^{-ikz\sin\theta} \sum_{k\neq j}^{N} \sum_{s=-\infty}^{\infty} \sum_{n=-\infty}^{\infty} (-i)^s e^{in\gamma_{jp}} G_{ks}^{jn} \times$$

$$\times J_n(k\mathbf{R}_{jp}\cos\theta n_o)\begin{pmatrix} a_s^k \\ b_s^k \end{pmatrix} \tag{7.23}$$

where

$$G_{ks}^{jn} = e^{in(s-n)\gamma_{kj}} H_{s-n}^2(k\mathbf{R}_{jk}\cos\theta n_o). \tag{7.24}$$

We substitute in (7.9) expression (7.15), (7.18) and (7.23) and then obtain [7]

$$\begin{pmatrix} U^j \\ V^j \end{pmatrix} = e^{-ikz\sin\theta} \sum_{k\neq j}^{N} \sum_{s=-\infty}^{\infty} \sum_{n=-\infty}^{\infty} (-i)^n e^{in\gamma_{jp}} \left[\begin{pmatrix} \alpha \\ 1-\alpha \end{pmatrix} e^{in\varphi} - \begin{pmatrix} a_s^k \\ b_s^k \end{pmatrix} G_{ks}^{jn} \right] \times$$

$$\tag{7.25}$$

$$\times J_n(k\mathbf{R}_{jp}\cos\theta n_o) - e^{-ikz\sin\theta} \sum_{k\neq j}^{N} \sum_{s=-\infty}^{\infty} \sum_{n=-\infty}^{\infty} (-i)^n e^{in\gamma_{jp}} \begin{pmatrix} a_n^j \\ b_n^j \end{pmatrix} H_n^2(k\mathbf{R}_{jp}\cos\theta n_o),$$

where G_{ks}^{jn} defined by the formula (7.24).

To find the unknown coefficients a_s^k, b_s^k, we must use the boundary conditions on the surface of each cylinder. These boundary conditions require the continuity of the tangential component of the electric and magnetic vectors on the surface of the cylinders. The use of the boundary conditions is analogous to [5] following a system of linear algebraic equations for finding the unknown coefficients a_s^k, b_s^k:

$$\sum_{k\neq j}^{N} \sum_{s=-\infty}^{\infty} \sum_{n=-\infty}^{\infty} \left[\left[\delta_{jk}\delta_{ns} + (1-\delta_{jk})G_{ks}^{jn}a_n^{j\,I} \right] a_s^k + (1-\delta_{jk})G_{ks}^{jn}a_n^{j\,II} b_s^k \right] =$$

$$= e^{-ikz\sin\theta} e^{in\theta} (\alpha a_n^{j\,I} + (1-\alpha)a_n^{j\,II}) \tag{7.26}$$

$$\sum_{k\neq j}^{N} \sum_{s=-\infty}^{\infty} \sum_{n=-\infty}^{\infty} \left[1 - \delta_{jk})G_{ks}^{jn}b_n^{j\,I}a_s^k + \left[(\delta_{jk}\delta_{ns} + (1-\delta_{jk})G_{ks}^{jn}b_n^{j\,II} \right] b_s^k \right] =$$

$$= e^{-ikz\sin\theta} e^{in\theta} (\alpha b_n^{j\,I} + (1-\alpha)b_n^{j\,II}) \tag{7.27}$$

or in the matrix form

$$
\begin{pmatrix} \delta_{jk}\delta_{ns} + (1-\delta_{jk})G_{ks}^{jn}a_n^{j\,I} & (1-\delta_{jk})G_{ks}^{jn}a_n^{j\,II} \\ (1-\delta_{jk})G_{ks}^{jn}b_n^{j\,I} & \delta_{jk}\delta_{ns} + (1-\delta_{jk})G_{ks}^{jn}b_n^{j\,II} \end{pmatrix} \begin{pmatrix} a_s^k \\ b_s^k \end{pmatrix} = \tag{7.28}
$$

$$
= e^{-ikz\sin\theta}\, e^{in\varphi} \begin{pmatrix} \alpha a_n^{j\,I} + (1-\alpha)a_n^{j\,II} \\ \alpha b_n^{j\,I} + (1-\alpha)b_n^{j\,II} \end{pmatrix}
$$

where δ_{jk}, δ_{ns} is Kronecker symbol.

The expressions for the components of the vector **E**, **H** can be found through the Hertz potentials U, V. The substitution of (7.18) with a glance (7.28) in (7.3)−(7.6) gives the corresponding relations longitudinal, azimuthal and radial component of electric and magnetic field.

$$
E_{(scat)R_{jp}} = -k\cos\theta n_o e^{-ikz\sin\theta} \sum_{j=1}^{N}\sum_{n=\infty}^{\infty}(-i)^n e^{in\gamma_{jp}}\left[\sin\theta\, H_n^{'(2)}a_n^j + \frac{i}{k\cos\theta R_{jp}}H_n^{(2)}b_n^j\right],
$$

$$
E_{(scat)\gamma_{jp}} = k\cos\theta n_o e^{-ikz\sin\theta} \sum_{j=1}^{N}\sum_{n=\infty}^{\infty}(-i)^n e^{in\gamma_{jp}}\left[-\frac{i}{k\cos\theta R_{jp}}\sin\theta\, H_n^{(2)}a_n^j + H_n^{'(2)}b_n^j\right],
$$

$$
E_{(scat)_z} = -ik\cos^2\theta n_o e^{-ikz\sin\theta} \sum_{j-1}^{N}\sum_{n=\infty}^{\infty}(-i)^n e^{in\gamma_{jp}}\left[H_n^{(2)}a_n^j\right], \tag{7.29}
$$

$$
H_{(scat)R_{jp}} = k\cos\theta n_o e^{ikz\sin\theta} \sum_{j=1}^{N}\sum_{n=\infty}^{\infty}(-i)^n e^{in\gamma_{jp}}\left[-\sin\theta\, H_n^{'(2)}a_n^j + \frac{i}{k\cos\theta R_{jp}}H_n^{(2)}b_n^j\right],
$$

$$
H_{(scat)\gamma_{jp}} = -k\cos\theta n_o e^{-ikz\sin\theta} \sum_{j=1}^{N}\sum_{n=\infty}^{\infty}(-i)^n e^{in\gamma_{jp}}\left[\frac{i}{k\cos\theta R_{jp}}\sin\theta\, H_n^{(2)}a_n^j - H_n^{'(2)}b_n^j\right],
$$

$$
H_{(scat)_z} = -ik\cos^2\theta n_o e^{-ikz\sin\theta} \sum_{j=1}^{N}\sum_{n=\infty}^{\infty}(-i)^n e^{in\gamma_{jp}}\left[H_n^{(2)}b_n^j\right]. \tag{7.30}
$$

7.3   Reflection of a Plane Wave from a Layer with the Fibrillar Structure

In this section, we consider the problem reflection of a plane wave from a layer with a slowly varying thickness. Consider the optical system, which consists of several areas with different refraction indices (the epidermis, the upper layer of the dermis with fibrillar structure, blood cells, and the lower layer of the dermis).

It should be noted that, to better match the real structure of the object under investigation, the interfaces between the layers are represented by wavy surface $z_i = H_i(x, y), i = \overline{1, 3}$.

Let a plane s- or p-polarized wave be incident to a layer at an angle θ. The reflected field must be found. We consider only the case of the p polarization.

We seek the reflected field in the form of waves with slowly varying amplitudes and quickly oscillating phases (see Chaps. 4, 5 and 6):

$$E_1 = \exp\left(\frac{i}{\epsilon}\tau_{inc}(\xi_1, \xi_2, \xi_3)\right) + \exp\left(\frac{i}{\epsilon}\tau_{1ref}(\xi_1, \xi_2, \xi_3)\right) A(\xi_1, \xi_2, \xi_3), \quad (7.31)$$

$$E_2 = \exp\left(\frac{i}{\epsilon}\tau_{2elap}(\xi_1, \xi_2, \xi_3)\right) B^+(\xi_1, \xi_2, \xi_3) +$$

$$+ \exp\left(\frac{i}{\epsilon}\tau_{3ref}(\xi_1, \xi_2, \xi_3)\right) B^-(\xi_1, \xi_2, \xi_3) \quad (7.32)$$

$$E_3 = \exp\left(\frac{i}{\epsilon}\tau_{3elap}(\xi_1, \xi_2, \xi_3)\right) C^+(\xi_1, \xi_2, \xi_3) +$$

$$+ \exp\left(\frac{i}{\epsilon}\tau_{4ref}(\xi_1, \xi_2, \xi_3)\right) C^-(\xi_1, \xi_2, \xi_3) + E_{scat}(\xi_1, \xi_2, \xi_3), \quad (7.33)$$

$$E_4 = \exp\left(\frac{i}{\epsilon}\tau_{4elap}(\xi_1, \xi_2, \xi_3)\right) D^+(\xi_1, \xi_2, \xi_3) +$$

$$+ \exp\left(\frac{i}{\epsilon}\tau_{5ref}(\xi_1, \xi_2, \xi_3)\right) D^-(\xi_1, \xi_2, \xi_3) + E_{\theta scat}(\xi_1, \xi_2, \xi_3), \quad (7.34)$$

$$E_5 = \exp\left(\frac{i}{\epsilon}\tau_{5elap}(\xi_1, \xi_2, \xi_3)\right) E(\xi_1, \xi_2, \xi_3), \quad (7.35)$$

where (7.33) is scattering on the parallel cylinders which simulate fibrillar structure and determined by the expression (7.29), $E_{\theta scat}$ is scattering on the inhomogeneous particles with an irregular shape, which simulate red blood cells and determined by expression (6.76), $\tau_{inc}, \tau_{1ref}, \tau_{2elap}, \tau_{3ref}, \tau_{3elap}, \tau_{4ref}, \tau_{4elap}, \tau_{5ref}, \tau_{5elap}$ are defined in Chap. 4. Amplitudes $A, B^\pm, C^\pm, D^\pm, E$ are sought in the form of series in powers of small parameters ϵ_x, ϵ_y, Note that the expression for the amplitudes is determined by the method described in Chap. 4.

The substitution of (7.31)–(7.35) in the boundary conditions of the type (4.6)–(4.11) generates the recursive system of equations. From this system, one can find the reflection coefficient in the principal approximation for the reflected field.

Let us briefly consider the problem of reflection of a Gaussian beam with an arbitrary cross section. This problem can be solved by expansion of counter propagating waves in terms of plane waves in the region of medium 1, their reflection by layer 2, and reverse transformation with a subsequent Huygens-Fresnel integral transformation to obtain the field in the initial section (see Chap. 4). The laser radiation intensity is defined by the form (4.69) of Chap. 4.

Substituting the expression (4.67) in (4.69) for the condition that the simulated layer is fibrillar structure we obtain the dependence of the laser radiation intensity for electrical parameters of the modeled biological structure.

7.4 Numerical Calculations for a Model Medium and Conclusions

Let us consider a model medium with the following characteristics: refractive indices of the layers are equal to $n_2^\circ = 1.50$, $n_3^\circ = 1.40$, $n_4^\circ = 1.35$, $n_5^\circ = 1.40$, the characteristic thicknesses of the layers amount to $d_2 = 65 \cdot 10^{-6}$, $d_3 = 565 \cdot 10^{-6}$, $d_4 = 90 \cdot 10^{-6}$, $n_1^\circ = 1$, $\chi_1 = 0$, $\chi_2 = \chi_3 = \chi_4 = \chi_5 = 10^{-5}$ and the following values of parameters $a_1 = -0.0024$, $b_1 = 0.020$, $a_2 = 0.021$, $b_2 = 0.030$, $a_3 = 0.041$, $b_3 = 0.051$, $c_1 = c_2 = c_3 = 10^{-2}$.

The values of parameters $a1$, $b1$, $a2$, $b2$, $a3$, $b3$, $c1$, $c2$, and $c3$ are chosen for the interface of each layer so that the surface shape are as close as possible to the interface shape of the corresponding layer in the structure of human skin, and the wavelength is $\lambda = 0.63$ μm (center of the line of a He–Ne laser).

The calculations were performed for monolayer particle spherulated modeling red blood cells, while the number of particles in the simulated layer is assumed to be ten, with the following parameters: the relative refractive index for the first five spherulated erythrocytes was assumed to be $1.035 + 10^{-5}i$, for others it was set as $1.033 + 10^{-5}i$ for a particle radius of 4.3 μm, number of cylinders modeling collagen fibers in the layer was assumed to be nine, the radius cylinder was assumed to be $1 \cdot 10^{-10}$, $1 \cdot 10^{-6}$, $2 \cdot 10^{-6}$, $1 \cdot 10^{-6}$, $1 \cdot 10^{-6}$, $3 \cdot 10^{-6}$, $1 \cdot 10^{-6}$, $1 \cdot 10^{-6}$, $2 \cdot 10^{-6}$. Note that in the numerical calculations we was considered the normal incidence of the electromagnetic wave and the reflected field was regarded in the main approximation.

Figure 7.2 shows the cross-section of the scattering field in the case of multiple scattering on the group, closely spaced cylinders, different radii in the far-field.

To consider scattering in the far field in expression (7.29), we replace of the Hankel function by its asymptotic representation for $kl \gg 1$:

$$H_n^2(kl) \sim \sqrt{\frac{2}{\pi kl}} e^{-i(kl-(2n+1)\pi/4)}, \quad l = R\cos\theta n_0, \gamma_{jp} \sim \gamma, \forall j.$$

Fig. 7.2 Cross-Section of the scattering field in the cylinder group with refractive index equal to 1.34 and *s*-polarization (**a**), *p*-polarization (**b**), not polarization (**c**)

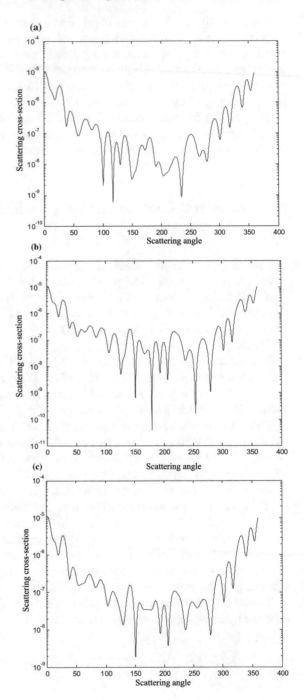

In this case, the scattering cross section is defined a

$$C_{scat} = \frac{W_{scat}}{I_i},$$

where I_i is incident intensity

$$W_{scat} = \int_A S_{scat} \cdot \mathbf{e_r} dA, \quad S_{scat} = \frac{c}{8\pi} \mathrm{Re}[\mathbf{E}^j_{scat} \times \mathbf{H}^{j*}_{scat}],$$

$$\mathbf{E}^j_{scat} = \frac{i}{k} \nabla \times \nabla \times (\mathbf{e_z} U^j_{scat}) + \nabla \times (\mathbf{e_z} V^j_{scat}),$$

$$\mathbf{H}^j_{scat} = -n_o \nabla \times (\mathbf{e_z} U^j_{scat}) + \frac{i}{k} \nabla \times \nabla \times (\mathbf{e_z} V^j_{scat}),$$

U^j_{scat}, V^j_{scat} defined by the expression (7.18).

Figure 7.3 shows the distribution of intensity radiation for absorbing multilayer and scatters light medium simulating human skin for specific electrical and geometrical characteristics of the simulated biological structure with the fibrillar structure.

Dependence of the intensity laser radiation on the coefficient refraction and absorption of the dermis with different electrical characteristics of the simulated tissue are shown in Fig. 7.4a, b.

It implies from the graph that with increasing absorption the simulated biological structure the intensity decreases consistent with the general theoretical concepts.

Thus, we can conclude that the model sufficiently sensitive to changes in electrical parameters, the simulated biological structure, in particular the coefficient absorption.

The model constructed allows variation of the optical parameters of the studied biological sample and the geometric characteristics, installing the relationship between them and the biological properties of the simulated tissue. Thus, by using this

Fig. 7.3 Intensity distribution for the modeled biological structure for specific values of the parameters and $\theta = 0°$, $\varphi = 0°$, $\psi = 0°$

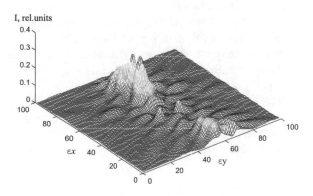

Fig. 7.4 Dependence of the intensity laser radiation on the coefficient refraction and absorption of the dermis with the absorption coefficient other layers assumed to be equal $\chi = 0.00001$ (**a**) and $\chi = 0.01$ (**b**)

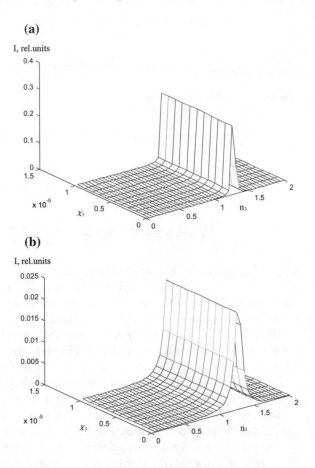

mathematical model we can measure spectral differences of normal and pathological tissue in the case of in vivo with the fibrillar structures for constructing a spectral autograph to assess determining pathological changes in the investigated biological samples, related to the change of electrophysical properties of the epidermis, the upper dermis and blood.

These dependences can be used to predict changes in the optical properties of the dermis, caused by therein various biophysical, biochemical and physiological processes, and they can also be calculated for lasers with different parameters, and well as the quantitative estimates which can be applied to processing and interpretation of experimental data.

The constructed model allows to determine not only the spectral distribution of the optical parameters of the biological environment associated with the light absorption in the upper layers the simulated biological fibrillar structure, but also the changes taking place under various factors that change functional and morphological condition of tissue and gives the possibility of simultaneous receipt of one plant aggregate results by varying the optical properties and characteristics the dimensions of the biological structure of the various structures.

References

1. D.A. Zimnyakov, Y.P. Sinichkin, O.V. Ushakov, Optical anisotropy of fibrous tissues. Quantum Electron. **37**(8), 777–783 (2007)
2. A.V. Papaev, The study of anisotropic optical properties and dynamics of immersion bleaching various tissues Ph.D. thesis. Saratov, 2007
3. A. Kienle, F. Forster, R. Hibst, Opt. Lett. **29**(22), 2617–2619 (2004)
4. Ch. Cantor, P. Schimel, Biophys. Chem. **1** (1984)
5. J. Schäfer, L. Siu-Chun, A. Kienle, Calculation of the near fields for the scattering of electromagnetic waves by multiple infinite cylinders at perpendicular incidence. JQSRT **113**, 2113–2123 (2012)
6. E.A. Ivanov, *Diffraction of Electromagnetic Waves on the Two Bodies* (Moscow, 1968)
7. Lee Siu-Chun, Scattering of polarized radiation by an arbitrary collection of closely spaced parallel nonhomogeneous tilted cylinders. J. Opt. Soc. Am. A **13**(11), 2256–2265 (1996)

Chapter 8
Theoretical Determination the Function of Size Distribution for Blood Cells

Abstract The mathematical model proposed for detection the the function of size distribution of form for blood cells. Using the mathematical model we can theoretically calculate the size distribution function for particles of irregular shape with a variety forms and structures of inclusions that simulate blood cells in the case of in vivo and determine the degree of aggregation, for example, the platelet for case in vivo, which may indicate the presence of pathogenesis.

8.1 Introduction

Regulation of erythrocytes' volume plays an important role in the life of an organism, and therefore the ratio of the surface area to the volume of these cells is an important parameter in determining the rheological properties of blood [1]. This makes it very important to develop methods that can effectively determine the globule distribution functions by size. Note that the blood test is one of the main tools of modern medical diagnostics. Erythrocytes (red blood cells) are blood's main cells, where they perform mainly transport and buffer functions. An important role in this process is played by the state of the cells themselves-their size, shape, deformability [2]. A change in the dispersion (width) of the distribution of cells size by 1% leads to an increase in the risk of mortality by 14% for patients with cardiovascular diseases [3]. Measurement of erythrocyte deformability gives additional medical information, especially important for the treatment of diseases such as sickle cell anemia, tropical malaria, diabetes mellitus, strokes, etc., [4]. Normal human erythrocytes have the form of a biconcave disk. Unlike most other cells, mammalian erythrocytes are devoid of nuclei. In addition, in contrast to platelets, also lacking a nucleus, red blood cells are more resistant to external influences [5], because their task is not activated in response to changes in the environment's composition, but to resistance to periodic deformations during circulation. All this explains why red blood cells are a common object of research. The development of methods for determining the properties of matter, in particular, the determination of distribution functions by globule size by scattering characteristics, is an important problem with which a number of biomedical, biophysical and geophysical problems are related. Cell volume is one of the

© Springer International Publishing AG, part of Springer Nature 2018   145
K. Kulikov and T. Koshlan, *Laser Interaction with Heterogeneous Biological Tissue*, Biological and Medical Physics, Biomedical Engineering, https://doi.org/10.1007/978-3-319-94114-1_8

main indicators of the functional and structural state of the cell [5]. Normally, a cell's volume regulation is performed by a number of interrelated physiological and biochemical processes [5, 6]. It is known that in some pathological conditions this regulation is disturbed [7–9]. In this chapter, a mathematical model has been developed to determine the distribution of blood cells by size. Normal human erythrocytes were chosen as the experimental model system for determining average sizes. From a mathematical point of view, the problem of reconstructing the globule distribution function in size and shape reduces to solving Fredholm integral equations of the first order. Note that in this approach we use the Tikhonov regularization method [10] and the use of a priori information on smoothness, non-negativity, and finiteness of the solution of the inverse problem. Most of the work on mathematical modeling in this area is devoted to solving an integral equation that relates an unknown distribution and a diffraction pattern. The kernel of this equation is a function describing the diffraction pattern corresponding to a single particle of a given shape. As the shape of a single particle, we will use a sphere or a cylinder [11]. For medical applications related to the operational diagnosis of erythrocytes, it is important that rapid analysis can be performed in the shortest possible time. Therefore, it is relevant to search for fairly simple but at the same time informative models in combination with analytical estimates of the basic parameters of unknown distributions that will allow us to apply new mathematical approaches to modeling without involving resource-intensive calculations.

In this chapter, we analyze geometrical characteristics of particles simulating erythrocytes in the upper layer of the dermis.

The problem consists of several steps. At the first stage, it is necessary to find the coefficient of reflection of a plane wave from a smoothly irregular layer simulating a given biological structure which consist of two continuous layers and the third layer containing inhomogeneous inclusions simulating blood cells with different refractive indices.

At the second stage, it is necessary to solve the problem of reflection of a with an arbitrary cross section for the above conditions (see Chap. 4). The construction of these parts is auxiliary.

At the third stage, we solve the problem of detection the of form for blood cells.

8.2 Reflection of a Plane Wave from a Layer with a Slowly Varying Thickness

For detection the the function of size distribution of form of blood, it is necessary to find the reflected field in the layer consisting irregularly shaped particles of various sizes and coefficients refraction.

It will be determined as follows: $\mathbf{E}_{blood} = \mathbf{E}_{ref} - \mathbf{E}_{skin}$, where $\mathbf{E}_{ref}$ is reflected field from all the simulated optical system $\mathbf{E}_{skin}$ is reflected field of layers: the epidermis, the upper layer of the dermis.

Let us consider the following optical scheme. The system consists of four regions with difference refractive indices(epidermis, the upper layer of the dermis, blood cells,the lower layer of the dermis) (see Fig. 6.2).

To attain the best agreement between the structure and the actual object under investigation, we represent the interfaces between the layers of the model medium in the form of certain surfaces $z_i = H_i(x, y), i = \overline{1, 3}$.

Let us suppose that a plane s- or p- polarized wave is incident on the layer at an angle θ. We consider only the case of the p polarization. We must find the reflected field.

We will seek the reflected field in the form of waves with slowly varying amplitudes and rapidly oscillating phases

$$E_1 = \exp\left(\frac{i}{\varepsilon}\tau_{inc}(\xi_1, \xi_2, \xi_3)\right) + \exp\left(\frac{i}{\varepsilon}\tau_{1ref}(\xi_1, \xi_2, \xi_3)\right) A(\xi_1, \xi_2, \xi_3, \varepsilon_x, \varepsilon_y),$$
$$(8.1)$$

$$E_2 = \exp\left(\frac{i}{\varepsilon}\tau_{2tr}(\xi_1, \xi_2, \xi_3)\right) B^+(\xi_1, \xi_2, \xi_3, \varepsilon_x, \varepsilon_y)+$$

$$+ \exp\left(\frac{i}{\varepsilon}\tau_{3ref}(\xi_1, \xi_2, \xi_3)\right) B^-(\xi_1, \xi_2, \xi_3, \varepsilon_x, \varepsilon_y),\qquad(8.2)$$

$$E_3 = \exp\left(\frac{i}{\varepsilon}\tau_{3elap}(\xi_1, \xi_2, \xi_3)\right) C^+(\xi_1, \xi_2, \xi_3, \varepsilon_x, \varepsilon_y)+$$

$$+ \exp\left(\frac{i}{\varepsilon}\tau_{4ref}(\xi_1, \xi_2, \xi_3)\right) C^-(\xi_1, \xi_2, \xi_3, \varepsilon_x, \varepsilon_y),\qquad(8.3)$$

$$E_4 = \exp\left(\frac{i}{\varepsilon}\tau_{4elap}(\xi_1, \xi_2, \xi_3)\right) D^+(\xi_1, \xi_2, \xi_3, \varepsilon_x, \varepsilon_y)+$$

$$\exp\left(\frac{i}{\varepsilon}\tau_{5ref}(\xi_1, \xi_2, \xi_3)\right) D^-(\xi_1, \xi_2, \xi_3, \varepsilon_x, \varepsilon_y)+$$

$$+ E_{4\theta scat}(\xi_1, \xi_2, \xi_3)\qquad(8.4)$$

$$E_5 = \exp\left(\frac{i}{\varepsilon}\tau_{5elap}(\xi_1, \xi_2, \xi_3)\right) E(\xi_1, \xi_2, \xi_3, \varepsilon_x, \varepsilon_y),\qquad(8.5)$$

where $E_{4\theta scat}$ is defined by formula (6.76) and $\tau_{1inc}, \tau_{1ref}, \tau_{2elap}, \tau_{3ref}, \tau_{3elap}, \tau_{4ref}, \tau_{4elap}$ are defined in Chap. 4. Amplitudes $A, B^\pm, C^\pm, D^\pm, E$ are sought in the form of power series in small parameter $\varepsilon_x, \varepsilon_y$, the expressions for the amplitudes are defined analogously to the method described in Chap. 4.

Substitution of expressions (8.1)–(8.5) into (4.6)–(4.11) generates a recurrence system of equations for reflected field from all the simulated optical system ($\mathbf{E}_{ref}$)

this system leads to reflection coefficient A and similar system of equations can be derived for reflected field of layers: the epidermis, the upper layer of the derms (E_{skin}), The expression for the reflection of a Gaussian beam with an arbitrary cross section is defined analogously to the method described in Chap. 4.

Determine the intensity:

$$I_{blood}(\theta, \lambda) = |E_{(blood)\perp}|^2 + |E_{(blood)\|}|^2,$$

where

$$E_{(blood)\perp} = \cos(\theta)E_{z_{blood}} + \sin(\theta)E_{x_{blood}},$$

$$E_{(blood)\|} = \sin(\theta)E_{z_{blood}} - \cos(\theta)E_{x_{blood}},$$

where E_x and E_z are given by the following expressions

$$\frac{\partial E_z}{\partial y} - \frac{\partial E_y}{\partial z} = -i\omega\mu_0\mu_j H_x, \quad \frac{\partial E_x}{\partial z} - \frac{\partial E_z}{\partial x} = -i\omega\mu_0\mu_j H_y, \quad (8.6)$$

$$\frac{\partial E_y}{\partial x} - \frac{\partial E_x}{\partial y} = -i\omega\mu_0\mu_j H_z, \quad \frac{\partial H_z}{\partial y} - \frac{\partial H_y}{\partial z} = i\omega\varepsilon_0\varepsilon_j E_x, \quad (8.7)$$

$$\frac{\partial H_x}{\partial z} - \frac{\partial H_z}{\partial x} = i\omega\varepsilon_0\varepsilon_j E_y, \quad \frac{\partial H_y}{\partial x} - \frac{\partial H_x}{\partial y} = i\omega\varepsilon_0\varepsilon_j E_z. \quad (8.8)$$

Formulas (8.6)–(8.8) correspond to the system of the Maxwell equations (4.3) in a Cartesian coordinate system. Thus, we obtained formulas allowing one to determine the explicit dependence of the intensity of laser radiation as a function of the refractive index and absorption coefficient for the system of blood vessels located in the upper dermis.

8.3 The Function of Size Distribution or Red Blood Cells

For defining the function of size distribution $\psi(\rho)$ we write the linear Fredholm integral equation the first kind

$$I_{blood}(\theta, \lambda) = \int_{\rho_{min}}^{\rho_{max}} s_i(\theta, \rho, \lambda)\psi(\rho)d\rho, \quad (8.9)$$

where $i = \overline{1, 5}$, ρ is radius of the particle, $I_{blood}(\theta, \lambda)$ coefficient scattering for a fixed angle θ, $s_i(\theta, \rho, \lambda)$ is kernel of the integral equation, which is defined as the scattering of light by individual non-spherical particles with irregular inclusion of modeling blood cells (erythrocytes, leukocytes, platelets, lipoproteins of low density and lipoproteins of high density). For the numerical determination of $f(\rho$ should be used Tikhonov regularization method [12].

8.3.1 Tikhonov Regularization Method

We consider the Fredholm integral equation of the first kind with smooth kernel $K(x, s)$

$$Au = \int_a^b K(x, s)u(s)ds = f(x), x \in [c, d] \tag{8.10}$$

where $f(x) = I_{blood}(\theta, \lambda)$, $K(x, s) = s_i(\theta, \rho, \lambda)$ $a \equiv \rho_{min}$ and $b \equiv \rho_{max}$, $u(s) = \psi(\rho)$.

We assume that $K(x, s)$ a real function that is continuous in the rectangle $G = ([c, d]) \times [a, b])$ and $f(x) \in L_2[c, d]$.

We also employ approximation $f_\delta(x)$ of function $f(x)$ such that $||f(x) - f_\delta(x)||_{L_2} \leq \delta$.

Based on the a priori assumptions, we suppose that $u(s)$ is a piecewise smooth function and choose $U = W_p^1[a, b]$. Let function K(x, s) be changed by function Kh(x, s), such that $||K(x, s) - K_h(x, s)||_{L_2(G)} \leq h$. Then, we have $||A - A_h||_{W_2^1 \rightarrow L_2} \leq h$, where A_h is an integral operator that corresponds to kernel $K_h(x, s)$.

Using the Tikhonov procedure for the construction of the regularization algorithm [12, 13], we proceed from expression (8.10) to the minimization of the smoothing functional

$$M^\alpha[u] = ||A_hu - f_\delta||_{L_2}^2 + \alpha||u||_{W_2^1}^2 \rightarrow \min, \tag{8.11}$$

where

$$||u||^2 = \int_a^b u^2(s)ds, \quad ||u'||^2 = \int_a^b (u'(s))^2 ds, \quad ||A_hu - f_\delta||^2 =$$

$$= \int_c^d \left[\int_a^b K(x, s)u(s)ds - f(x)\right]^2 dx,$$

Then, expression (8.11) is represented as

$$M^\alpha[u] = \int_c^d \left[\int_a^b K(x, s)u(s)ds - f(x)\right]^2 dx +$$

$$+ \alpha \left[\int_a^b u^2(s)ds + \int_a^b (u'(s))^2 ds\right] \rightarrow \min. \tag{8.12}$$

The Tikhonov condition [12, 13] follows from condition (8.12):

$$(A_h^* A_h + \alpha C) u^\alpha = A_h^* f$$

Here, A_h is the operator from $W_2^1[a, b]$ $L_2[c, d]$, A_h^* is the conjugate operator with respect to A_h, A_h^* is the operator from $L_2[c, d]$ $W_2^1[a, b]$, and C is the operator the matrix of which is determined in [12, 13].

In the above formulation, we consider operator A_h of the original integral equation that acts from $L_2[a, b]$ to $L_2[c, d]$ (i.e., the information regarding the smoothness of the exact solution is missing). Then, the smoothing functional is written as

$$M^\alpha[u] = ||A_h u - f_\delta||_{L_2}^2 + \alpha ||u||_{L_2}^2 \to \min$$

and the Tikhonov equation is represented as

$$(A_h^* A_h + \alpha E) u^\alpha = A_h^* f,$$

where E is the unity operator.

Note that function u^α that minimizes functional (8.11) or (8.12) depends on regularization parameter α. To determine the regularization parameter, we employ the method of relative residual

$$\frac{||A u^\alpha - f||}{f} = \delta \tag{8.13}$$

Thus, expression (8.13) makes it possible to automatically determine the regularization parameter.

8.4 Numerical Calculations for a Model Medium and Conclusions

Let us consider a model medium with the following characteristics. Typical layer thicknesses are equal to $d_2 = 65 \cdot 10^{-6}$, $d_3 = 565 \cdot 10^{-6}$, $d_4 = 90 \cdot 10^{-6}$, $n_1^\circ = 1$, $\chi_1 = 0$, $\chi_2 = \chi_3 = \chi_4 = \chi_5 = 10^{-5}$, refractive indices of the layers are $n_2^\circ = 1.50$, $n_3^\circ = 1.40$, $n_4^\circ = 1.35$, $n_5^\circ = 1.40$ and the following values of parameters are $a_1 = -0.0024$, $b_1 = 0.020$, $a_2 = 0.021$, $b_2 = 0.030$, $a_3 = 0.041$, $b_3 = 0.051$, $c_1 = c_2 = c_3 = 10^{-2}$. The values of parameters for the interfaces between the layers are chosen so that the shape of the surface is maximally close to the shape of the boundary of the corresponding layer in the structure of the normal human dermis, and wavelength is $\lambda = 0.63\,\mu m$ (center of the line of a He–Ne laser).

Since the erythrocyte contains no cell organelles, its cellular membrane is very thin and does not noticeably affect the scattering of light; consequently, the erythrocyte can be treated as a homogeneous scatterer. Thus, our computations were performed for monolayer spherulated particles simulating erythrocytes; the number of particles

Fig. 8.1 Function of size distribution for red blood cells

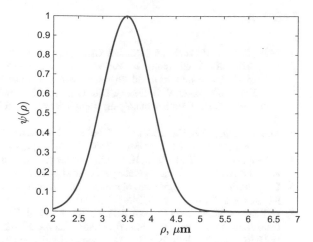

in the layer being simulated was assumed to be ten for the following parameters: the relative refractive index for the first five spherulated erythrocytes was assumed to be $1.035 + 10^{-5}i$; for the remaining erythrocytes, it was set as $1.033 + 10^{-5}i$, for a particle radius of 4.3 μm. All computations were performed up to 32 decimal places.

Figure 8.1 show the function of size distribution for blood corpuscle (erythrocyte). Based on the mathematical model (8.9) we can theoretically calculate the size distribution function for particles of irregular shape with a variety forms and structures of inclusions that simulate blood cells in the case of in vivo and determine the degree of aggregation, for example, the platelet for case in vivo, which may indicate the presence of pathogenesis.

We have described the mathematical model for calculating the function of size distribution for blood corpuscle of propagation of light in a multilayer biotissue in the case of the interaction with noncoagulating laser radiation. The model was implemented in the form of a software package, which makes it possible to vary automatically the composition of biological objects, their electrophysical parameters, characteristic thicknesses of layers, as well as characteristic sizes of various biological structures under investigation on the same setup for recording the dependence between these parameters. This makes the software developed here an effective and convenient tool for investigations in biomedical optics.

In this chapter, a mathematical model has been developed that allows one to calculate the dispersion and the particle size distribution function from the experimental data, taking into account the different models of light scattering by particles. The model allows one to dynamically change the geometry of an individual particle, graphically visualize the work of numerical methods of regularizing inverse problems and compare their work with the use of a priori dependencies and estimates, which is very important for the application of the results obtained in medical practice.

References

1. M.R. Clark, N. Mohandas, S.B. Shohet, Osmotic gradient ektacytometry: comprehensive characterization of red cell volume and surface maintenance. Blood **61**, 899–910 (1983)
2. G.I. Kozinets, The study of the blood system in clinical practice. M. Triada-X **12**, 480c (1997)
3. K.V. Patel, L. Ferrucci, W.B. Ershler, D.L. Longo, J.M. Guralnik, Red blood cell distribution width and the risk of death in middle-aged and older adult. Arch. Intern. Med. **169**(5), 515–523 (2009)
4. J.G.G. Dobbe, M.R. Hardeman, G.J. Streekstra, J. Strackee, C. Incc, C.A. Grimbergen, Analyzing red blood cell-deformability distributions. Blood Cells Mol. Dis. **28**(3), 373–384 (2002)
5. F. Lang, G.L. Busch, M. Ritter, H. Volkl, S. Waldegger, E. Gulbins, D. Haussinger, Functional significance of cell volume regulatory mechanisms. Physiol. Rev. **78**, 247–306 (1998)
6. E.K. Hoffmann, L.O. Simonsen, Membrane mechanisms in volume and pH regulation in vertebrate cells. Physiol. Rev. **69**, 315–382 (1989)
7. A. Iolascon, Miraglia del Giudice, S. Perrotta, L. Morle, J. Delaunay, Hereditary spherocytosis: from clinical to molecular defects. Haematologica **83**, 240–257 (1998)
8. W.H. Reinhart, The influence of extracorporeal circulation on erythrocytes and flow properties of blood. J. Thorac. Cardiovasc. Surg. **100**(4), 538–543 (1990)
9. R. Hallgren, K. Svenson, E. Johansson, U. Lindh, Abnormal calcium and magnesium stores in erythrocytes and granulocytes from patients with inflammatory connective tissue diseases. Relationship to inflammatory activity and effect of corticosteroid therapy. Arthritis Rheum **28**, 169–173 (1985)
10. A.V. Goncharsky, A.S. Leonov, A.G. Yagola, The generalized residual principle. J. Comput. Math. Math. Phys. **13**(2), 294–302 (1973)
11. J.B. Riley, Y.C. Agrawal, Sampling and inversion of data in diffraction particle sizing. Appl. Opt. **30**(33), 4800–4817 (1991)
12. A.N. Tikhonov, V.Y. Arsenin, *Solutions of Ill- Posed Problems* (Halsted, New York, 1977)
13. A.N. Tikhonov, A.V. Goncharskii, V.V. Stepanov, A.G. Yagola, *Numerical Methods for Solution of Ill- Posed Problems* (Kluver Academic, Dordrecht, 1995)

Chapter 9
Study of Optical Properties of Biotissues by the Intracavity Laser Spectroscopy Method

Abstract We describe a mathematic model for predicting the absorption spectrum and dispersion of a section of a biological structure consisting of epidermis, upper layer of the derma, blood, and lower layer of the derma and placed in the cavity of an optical resonator. It should be noted that the biological structure was represented by layers with different optical and geometrical parameters illuminated by a laser beam.

9.1 Introduction

Optical methods (including traditional optical spectroscopy) based on analysis of reflection, transmission, and fluorescence spectra of biological tissues play an important role among modern physical methods of analysis in biology and medicine. The most effective methods that make it possible to study processes in complex biological systems are optical intracavity techniques. The application of intracavity laser spectroscopy makes it possible to obtain more exact estimates of optical parameters of the medium, which cannot be detected by conventional methods. Optical methods make it possible to analyze processes without violating (modifying) living structures in complex biosystems. However, the application of these methods requires the development of appropriate mathematical models for better understanding the process of interaction of a laser beam with a biological object and for extending potentialities, reliability, and availability of optical technologies, which would make it possible to theoretically predict electrophysical parameters as characteristics of the structural state of biological tissues (including human derma). The determination of optical indices of a biological tissue is a complicated problem due to the complex and heterogeneous structure of the tissue itself. Modern techniques for determining optical parameters of biosystems involve the solution of the inverse scattering problem for various theoretical models such as the Monte Carlo method [1, 8], diffusion approximation [2–4], and KubelkaMunk method of flow models [5–7].

In this study, mathematical model is constructed, which makes it possible to vary electrophysical and geometrical parameters (layer thickness) of the section of a biological tissue being modeled and to represent the result in the form of a graph

© Springer International Publishing AG, part of Springer Nature 2018   153
K. Kulikov and T. Koshlan, *Laser Interaction with Heterogeneous Biological Tissue*, Biological and Medical Physics, Biomedical Engineering,
https://doi.org/10.1007/978-3-319-94114-1_9

describing the dependence of the real and imaginary parts of the refractive index of the model structure on the wavelength (dispersion curves and absorption spectra) for each version of calculations.

The problem includes several stages. At the first stage, the reflectance of a plane wave from a smoothly irregular layer simulating a given biological structure must be determined (see Chap. 4).

At the second stage, we must solve the problem of reflection of a Gaussian beam with an arbitrary cross section from a smoothly irregular layer simulating the given biological structure. The problem is solved by expanding the fields of counterpropagating waves in plane waves in region 1 of the medium and their reflection from layer 2 and carrying out inverse transformation followed by the Huygens–Fresnel integral transformation to obtain the field in the initial reference cross section after the circumvention of the cavity (see Chap. 4). The constructions at these stages are auxiliary.

We consider here natural oscillations of a linear resonator loaded with a layer modeling a given biological structure. The constructions are based on solving auxiliary problems of the first and second stages.

Chapter is based on the results of the [9, 10].

9.2 Integral Equation for Natural Oscillations of Field in a Resonator

Let a cell with a sample of a biological tissue (tissue section) be located in the vicinity of the Z axis in domain Ω of the cavity.

Since natural oscillations in ring and linear resonators are retuned in different ways upon the introduction of inhomogeneities in the cavity, we will consider for definiteness the simpler case of a linear resonator. We can write the integral equation

$$\Phi(\xi_1') = \gamma \int_{-\infty}^{\infty} K_1(\xi_1', \xi_1'') E_{ref}(\Phi(\xi_1'', \xi_2'')) d\xi_1'', \qquad (9.1)$$

where $E_{ref}(\Phi(\xi_1'', \xi_2''))$ is a linear combination of $\Phi(\xi_1'', \xi_2'')$ and its derivatives and is defined in the Chap. 4, and $K_1(\xi_1', \xi_1'')$ is the kernel of the integral transformation of the field,

$$K_1(\xi_1', \xi_1'') = \sqrt{\frac{k}{2\pi i B}} e^{\frac{ik}{2B}(A\xi_1''^2 + D\xi_1'^2 - 2\xi_1'\xi_1'') + ikL},$$

It should be noted that a characteristic feature of (9.1) is the presence of the derivative of function $\Phi(\xi_1')$ in the integrand. We will seek Φ and γ in the form of a power expansion in small parameter ε characterizing the smoothness of variations in the properties of the medium over a wavelength; i.e.,

$$\Phi = \psi_0 + \varepsilon\varphi_{01} + O(\varepsilon^2) \qquad (9.2)$$

$$\gamma = \psi_1 + \varepsilon\varphi_{11} + O(\varepsilon^2) \tag{9.3}$$

We substitute expansions (9.2) and (9.3) into the integral equation (9.1). Then, in the main approximation, we obtain

$$\psi_0^\pm(\xi_1') = \psi_1\left[\int_{-\infty}^\infty K_1(\xi_1',\xi_1'')S_{00}(\xi_1'',\xi_2'')\psi_0(\xi_1'')d\xi_1''\right] \tag{9.4}$$

Multiplying (9.1) by $\psi_1 S_{00}(\xi_1',\xi_2')\psi_0^\pm(\xi_1')$, integrating with respect to ξ_1', and taking into account the main approximation of (9.4), we obtain the following corrections to eigenvalues:

$$\varphi_{11} = \pm\psi_1\left[\int_{-\infty}^\infty \psi_0^+\psi_0^- S_x(\xi_1'',\xi_2'')d\xi_1'' - \int_{-\infty}^\infty \psi_0^\pm\frac{\partial S_x(\xi_1'',\xi_2'')}{\partial k_x}\frac{\partial\psi_0^\mp}{\partial\xi_1'}d\xi_1''\right]\Delta^{-1},$$

$$\Delta = \left[\int_{-\infty}^\infty \psi_0^+\psi_0^- S_{00}(\xi_1'',\xi_2'')d\xi_1''\right],$$

where

$$S_{00}(\xi_1'',\xi_2'') = \frac{A_{oo}^\frown(\xi_1''^\sim + \xi_2''^\sim, k_{1y}, k_{1x})}{\alpha}, \tag{9.5}$$

$$S_x(\xi_1'',\xi_2'') = \frac{1}{\alpha}\left[A_{1o}^\frown(\xi_1''^\sim + \xi_2''^\sim, k_{1y}, k_{1x}) + \frac{k_{13}}{kn_1}\xi_1'' A_{0000}(\xi_1''^\sim + \xi_2''^\sim, k_{1y}, k_{1x})\right],$$

$$\frac{\partial S_x(\xi_1'',\xi_2'')}{\partial k_x} = \left[\frac{k_x^o}{ikn_1\alpha}\left[\frac{\partial A_{oo}^\frown(\xi_1''^\sim + \xi_2''^\sim, k_{1y}, k_{1x})}{\partial k_{1x}} + \frac{\partial A_{oo}^\frown(\xi_1''^\sim + \xi_2''^\sim, k_{1y}, k_{1x})}{\partial k_{1y}}\right]\right],$$

quantities α, k_{13}, k_x^o, $A_{oo}^\frown(\xi_1''^\sim + \xi_2''^\sim, k_{1y}, k_{1x})$, $A_{1o}^\frown(\xi_1''^\sim + \xi_2''^\sim, k_{1y}, k_{1x})$ and $A_{0000}(\xi_1''^\sim + \xi_2''^\sim, k_{1y}, k_{1x})$ arc defined in Chap. 4.

The solution to (9.4) was sought in the form of a power series expansion in the eigenfunctions of an ideal resonator:

$$\psi_0^\pm = \sum_n \tilde{a}_n E_n^\pm(\xi_1''), \tag{9.6}$$

where field $E_n^\pm(\xi_1'')$ can be represented as the sum of counterpropagating waves

$$E_n^\pm(\xi_1'') = E_n^+(\xi_1'') + E_n^-(\xi_1''),$$

$$E_n^+(\xi_1'') = C_n H_n\left(\frac{\xi_1'' \cdot \sqrt{2}}{\omega}\right)\exp\left(-i(n+1/2)g + ikL + \frac{i\xi_1''^2}{q^+}\right),$$

$$E_n^-(\xi_1'') = C_n H_n\left(\frac{\xi_1'' \cdot \sqrt{2}}{\omega}\right) \exp\left(i(n+1/2)g - ikL - \frac{i\xi_1''^2}{q^-}\right).$$

Let us write the matrix equation for determining coefficients $\widetilde{a}_n$

$$\widetilde{a}_m = \psi_1 \sum_n \widetilde{a}_{mn}\widetilde{a}_n, \tag{9.7}$$

where

$$\widetilde{a}_{mn} = e^{-i(n+1/2)g}e^{-i(m+1/2)g}\int_{-\infty}^{\infty} C_m C_n e^{-\xi_1''^2} H_n(\xi_1'')H_m(\xi_1'')S_{00}(\xi_1'',\xi_2'')d\xi_1'',$$

$$g = \arccos\left[\frac{A+D}{2}\right], \quad C_n = \sqrt{\frac{1}{2^n n!\omega\pi}}, \quad C_m = \sqrt{\frac{1}{2^m m!\omega\pi}}, \quad \omega = \sqrt{\frac{\sin g}{B}},$$

$$\frac{1}{q} = \left[\frac{A+D}{2} + i\sqrt{1 - \frac{(A+D)^2}{4}} - A\right](2B)^{-1}$$

A, B and D are the elements of the wave matrix of the resonator; L is the resonator length; H_n, H_m are Hermitean polynomials; $k = 2\pi/\lambda$ is the wavenumber and $S_{00}(\xi_1'',\xi_2'')$ is defined by expression (9.5).

Matrix system (9.7) is a system of homogeneous linear algebraic equations, which is used for determining the transverse modes of the resonator by formula (9.6) after the calculation of eigenvectors, while the eigenfrequencies of these modes can be found from the equality of the determinant of this system to zero. Thus, at this stage, the frequencies of natural oscillations of the optical resonator loaded with the sample of the biological tissue under investigation were connected by formula (9.7) with the electrophysical parameters of this biological structure, such as the real and imaginary parts of their refractive indices and sizes.

Further testing and analysis of the above dependences will be carried out using numerical methods.

9.3 Numerical Calculations for a Model Medium and Conclusions

Let us consider an optical resonator with a model medium which has the following parameters: the distance L between the mirrors is 11 cm; radii of mirrors M_1 and M_2 are 100.0 and 46.3 cm, respectively.

It should be noted that, for better matching to the real structure of the object under investigation, the interfaces between the layers are represented by wavy surface $z_1 = H_1(x, y)$, $z_2 = H_2(x, y)$, $z_3 = H_3(x, y)$, where $H_1(x, y) = c_1 \sin(a_1 x +$

$b_1 y), H_2(x, y) = c_2 \sin(a_2 x + b_2 y), H_3(x, y) = c_3 \sin(a_3 x + b_3 y), c_1, a_1, b_1, c_2, a_2,$
b_2, c_3, a_3, b_3 are some arbitrary constants. The arbitrarily preset constants are:
$a_1 = -0.0024, b_1 = 0.020, a_2 = 0.021, b_2 = 0.030, a_3 = 0.041, b_3 = 0.051, c_1 =$
$c_2 = c_3 = 10^{-2}$. The values of parameters $a_1, b_1, a_2, b_2, a_3, b_3, c_1, c_2, c_3$ are selected
for the interfaces between the layers so that the shape of the surface matches as close
as possible to the interface between the corresponding layers in the structure of the
normal human derma. All calculations were carried out for the principal transverse
mode.

Figure 9.1a, b show the dependence of the imaginary part of the refractive index
(absorptance) of the epidermis on the wavelength. It can be seen from the curves
that the refractive index of the epidermis in the ultraviolet range is high. This is
apparently due to the fact that, at a given wavelength, light in the surface layer is
strongly absorbed, mainly by melanin.

The dependence of the real part of the refractive index of the epidermis on the
wavelength is shown in Fig. 9.2a, b. It can be seen from Fig. 9 that the maximal values
of the real part of the refractive index of the epidermis are attained for wavelengths
for which the values of the refractive index of the epidermis are minimal and vice
versa, which is in conformity with the general theoretical concepts. It should be
noted that the mathematical model constructed here is quite sensitive to change in
the optical parameters of the model medium and that the ranges of quantities n_2 (real
part of the refractive index of the epidermis) and χ_2 (imaginary part of the refractive
index of the epidermis) are close to experimental values of the complex refractive
index for the biological structure being modeled that were obtained without using
the intracavity model [11].

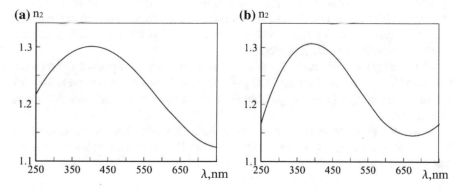

Fig. 9.1 a Dependence of the real part of the refractive index of the epidermis on wavelength
for the following parameters of the model medium: the imaginary part of the refractive index of
the epidermis is 0.00001, he refractive index of the upper derma is $1.3 + 0.00001i$, the refractive
index of blood is $1.3509 + 0.00001i$, the refractive index of the lower derma is $1.3 + 0.00001i$, the
thicknesses of the epidermis, upper derma, and blood are $64, 600$ and $80\,\mu m$, **b** Dependence of the
real part of the refractive index of the epidermis on wavelength for the following parameters of the
model medium: the imaginary part of the refractive index of the epidermis is 0.00001, he refractive
index of the upper derma is $1.3 + 0.00001i$, the refractive index of blood is $1.35 + 0.00001i$, the
refractive index of the lower derma is $1.45 + 0.00001i$, the thicknesses of the epidermis, upper
derma, and blood are $65, 80$ and $600\,\mu m$, respectively

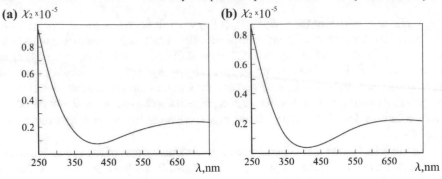

Fig. 9.2 **a** Dependence of the imaginary part of the refractive index of the epidermis on wavelength for the following parameters of the model medium: the real part of the refractive index of the epidermis is 1.3, the refractive index of the upper derma is $1.33 + 0.00001i$, the refractive index of blood is $1.35 + 0.00001i$, the refractive index of the lower derma is $1.45 + 0.00001i$, the thicknesses of the epidermis, upper derma, and blood are 65, 600 and 80 μm, respectively. **b** Dependence of the imaginary part of the refractive index of the epidermis on wavelength for the following parameters of the model medium: the real part of the refractive index of the epidermis is 1.3, the refractive index of the upper derma is $1.33 + 0.00001i$, the refractive index of blood is $1.3501 + 0.00001i$, the refractive index of the lower derma is $1.45 + 0.00001i$, the thicknesses of the epidermis, upper derma, and blood are 65 μm, 600 μm and 80 μm, respectively

The model constructed here makes it possible to determine not only the spectral distributions of optical parameters of a biological medium, which are associated with absorption of light in the upper layers of the biological structure being simulated, but also their variations occurring under the action of various factors leading to a change in the functional and morphological state of the biological tissue. The model also makes it possible to obtain simultaneously on the same setup an aggregate of results of variation of electrophysical parameters and characteristic sizes of various biological structures under investigation.

Thus, using the mathematical model constructed here, it is possible to measure the spectral differences in normal and pathological tissues in vitro for constructing a spectral autograph to assess pathological changes in biological samples under investigation.

Analogous dependences can be calculated for lasers with other parameters and can be used for processing experimental dispersion and absorption curves for biological tissues.

References

1. J. Qu, C. MacAulay, S. Lam et al., Laser-induced fluorescence spectroscopy at endoscopy: tissue optics, Monte Carlo modeling, and in vivo measurements. Opt. Eng. **34**(11), 3334–3343 (1995)
2. R.A.J. Groenhuis, H.A. Ferverda, J.J. Ten Bosch, Scattering and absorption of turbid materials determined from reflection measurements 1: Theory. Appl. Opt. **22**(16), 2456–2462 (1983)
3. J.L. Karagiannes, Z. Zhang, B. Grossweiner et al., Applications of the 1-D diffusion approximation to the optics of tissues and tissue phantoms. Appl. Optics. **28**(12), 2311–2317 (1989)

4. D.J. Maitland, J.T. Walsh, J.B. Prystowsky, Optical properties of human gallbladder tissue and bile. Appl. Opt. **32**(4), 586–591 (1993)
5. M.J.C. Van Gemert, G.A.C. Schets, M.S. Bishop et al., Optics of tissue in a multi-layer slab geometry. Laser Life Sci. **1**(2), 1–18 (1988)
6. M.J.C. Van Gemert, S.L. Jacques, H.J.C.M. Sterenborg et al., Skin optics. IEEE J. Biomed. Eng. **36**(12), 1146–1154 (1989)
7. K.M. Giryayev, N.A. Ashurbekov, O.V. Kobzev, Optical studies of biological tissues: determination of the scattering and absorption coefficients. Lett. J. Tech. Phys. **21**, 48–52 (2003)
8. V.V. Tuchin, S.R. Utz, I.V. Yaroslavskii, Tissue optics, light distribution, and spectroscopy. Opt. Eng. **33**, 3178–3176 (1994)
9. K.G. Kulikov, Study of electrophysical characteristics of blood formed elements using intracavity laser spectroscopy. I. Simulation of light scattering by an ensemble of biological cells with complicated structures. Tech. Phys. **59**(4), 576–587 (2014)
10. K.G. Kulikov, Simulation of electrophysical properties of biological tissues by the intracavity laser spectroscopy method. Tech. Phys. **54**(3), 435–439 (2009)
11. V.V. Tuchin, *Lasers and Fiber Optics in Biomedical Studies* (Saratovsky Univ, Saratov, 1998)

Chapter 10
Study of the Optical Characteristics of Thin Layer of the Biological Sample

Abstract We construct the mathematical model, which makes it possible to vary the characteristic sizes of roughness, the electrophysical parameters of the biological sample under investigation, and its geometrical characteristics and to establish the relations between these parameters and biological properties of the biological tissue being modeled, as well as to calculate theoretically the absorption spectra of optically thin biological samples placed into the cavity of an optical resonator.

10.1 Introduction

Most biological surfaces are rough to a certain extent. The roughness of the surface affects the characteristics of wave propagation and scattering (namely, the characteristics of a wave propagating over such a surface differ from analogous characteristics in the case of propagation over a smooth surface). A wave incident on a rough surface not only reflects specularly, but is also scattered in all other directions. In analysis of scattering from a rough surface, the extent of roughness of the surface is connected with the wavelength of incident radiation and depends on the direction of wave propagation and scattering. In this connection, it is important to study the effect of roughness on the spectral characteristics of the biological structure being simulated.

It should be noted that using a resonator, it is possible to obtain more exact estimates of optical parameters of the medium taking into account the roughness, which cannot be detected using conventional methods. Thus, it is expedient to consider the problem of natural oscillations of a linear resonator loaded with an optically thin layer simulating a certain biological structure. The biological structure is represented by an optically thin layer with certain optical and geometrical characteristics, which is illuminated by a laser beam.

The problem includes the following three consecutive stages. At the first stage, the problem of scattering from the rough boundary had to be solved and the coefficient of reflection of a plane wave from a smoothly irregular layer must be determined taking into account the roughness of the boundary simulating the given biological medium. At the second stage, the problem of reflection of a Gaussian beam with an arbitrary cross section had to be solved. The problem was solved by expanding the fields

© Springer International Publishing AG, part of Springer Nature 2018   161
K. Kulikov and T. Koshlan, *Laser Interaction with Heterogeneous Biological Tissue*, Biological and Medical Physics, Biomedical Engineering, https://doi.org/10.1007/978-3-319-94114-1_10

of counterpropagating waves in plane waves in the region of medium 1 and their reflection from layer 2 using inverse transformation followed by HuygensFresnel integral transformation to obtain the field in the initial reference cross section after the circumvention of the resonator (see Chap. 4). At the third stage, the effect of roughness on the spectral characteristics of the biological sample being simulated was investigated.

Chapter is based on the results of the [1, 2].

10.2   Scattering of a Plane Wave from a Rough Surface

As noted above, the surfaces of real bodies (in particular, in biology) are not always perfectly smooth to a certain extent; for this reason, reflection and refraction of waves from such surfaces are accompanied by phenomena which are not observed in the case of perfectly smooth interfaces. The form of scattering from a rough surface is determined by the set of the following factor: the degree of smoothness is determined by the relation between the wavelength of incident radiation and the geometrical parameters of the surface; the polarization of the primary wave as well as the reflecting and refracting properties of the substance also play a significant role. Rigorous methods for solving problems in the case of a rough surface do not exist.

The problem can be solved only approximately under certain constraints imposed on the size and shape of roughness. The scattered field is calculated using the method of small perturbations and the Kirchhoff method. In this study, we are using the small perturbation method for calculating the scattered field.

To apply the small perturbation method correctly, we assume that roughness of the surface under investigation is small and gently sloping on the wavelength scale are small and gently sloping on the wavelength scale. The slope of roughness indicates [3] that the inclination of the surface is small on the average; i.e., $\sigma_H^2/l_H^2 \ll 1$, where $\sigma_H^2 \equiv \langle H^2 \rangle$ is the standard deviation from the unperturbed surface $z = 0$ and l_H is the characteristic size of irregularities. The smallness of irregularities means that moments $\langle H^m \rangle$ are small as compared to the relevant powers of the wavelength, $\langle H^m \rangle \ll \lambda^m$; in particular, $\sigma_H^2 \ll \lambda^2$. As a result, for small and gently sloping irregularities, we can use the expansion of the boundary conditions as well as the sought solutions into a power series in small parameters $H/\lambda \ll 1$ and $\sigma_H/l_H \ll 1$ (i.e., we apply the perturbation method). Let us suppose that a plane monochromatic of unit amplitude is incident on a rough surface. We consider two media with refractive indices n_1 and n_2. The equation of the surface has the form $z = H(x, y)$; we assume that $\left|\frac{\partial H}{\partial x}\right| \ll 1, \left|\frac{\partial H}{\partial y}\right| \ll 1$.

We denote by E_1 and E_2 the amplitudes of the electric field in the upper and lower media, respectively. The electric field amplitude E_1 in the upper medium satisfies the equation

$$\frac{\partial^2 E_1}{\partial x^2} + \frac{\partial^2 E_1}{\partial y^2} + \frac{\partial^2 E_1}{\partial z^2} + k^2 n_1^2 E_1 = 0 \qquad (10.1)$$

while electric field amplitude E_2 in the lower medium satisfies the equation

$$\frac{\partial^2 E_2}{\partial x^2} + \frac{\partial^2 E_2}{\partial y_2^2} + \frac{\partial^2 E_2}{\partial z^2} + k^2 n_2^2 E_2 = 0 \tag{10.2}$$

where k is the wavevector and n_j is the complex refractive index $n_j = n_j^o + i\chi_j$, $j = \overline{1,2}$ with the boundary conditions in the form

$$E_1|_{z=H(x,y)} = E_2|_{z=H(x,y)}, \tag{10.3}$$

$$\frac{1}{n_1^2}\frac{\partial E_1}{\partial \mathbf{n}}\Big|_{z=H(x,y)} = \frac{1}{n_2^2}\frac{\partial E_2}{\partial \mathbf{n}}\Big|_{z=H(x,y)}, \tag{10.4}$$

where $\mathbf{n}$ is the unit vector of the outward normal with the following components:

$$\mathbf{n} = \left(\alpha\frac{\partial H}{\partial x}, \alpha\frac{\partial H}{\partial y}, -\alpha\right), \quad \alpha = \frac{1}{\sqrt{(1 + (\frac{\partial H}{\partial x})^2 + (\frac{\partial H}{\partial y})^2}}.$$

We must find the reflected field taking into account the roughness of the interface between the media. We consider only the case of the p polarization. We expand boundary condition (10.3) into a power series in H:

$$(E_1)\,|_{z=0} + H\left(\frac{\partial E_1}{\partial z}\right)_{z=0} + \frac{H^2}{2}\left(\frac{\partial^2 E_1}{\partial z^2}\right)_{z=0} + \cdots.$$

$$= (E_2)\,|_{z=0} + H\left(\frac{\partial E_2}{\partial z}\right)_{z=0} + \frac{H^2}{2}\left(\frac{\partial^2 E_2}{\partial z^2}\right)_{z=0} + \cdots \tag{10.5}$$

Let us consider the boundary condition of type (10.4).

$$\frac{\partial E_1}{\partial \mathbf{n}}\Big|_{z=H(x,y)} = \frac{1}{n_1^2}\frac{\partial E_1}{\partial \mathbf{n}}\Big|_{z=H(x,y)} = \frac{1}{n_1^2}\left(n_x\frac{\partial E_1}{\partial x} + n_y\frac{\partial E_1}{\partial y} + n_z\frac{\partial E_1}{\partial z}\right) =$$

$$= \frac{1}{n_1^2}\left(\alpha\frac{\partial H}{\partial x} + \cdots\right)\frac{\partial}{\partial x}\left(E_1|_{z=0} + H\frac{\partial E_1}{\partial z}|_{z=0} + \cdots\right) + \tag{10.6}$$

$$+ \frac{1}{n_1^2}\left(\alpha\frac{\partial H}{\partial y} + \cdots\right)\frac{\partial}{\partial y}\left(E_1|_{z=0} + H\frac{\partial E_1}{\partial z}|_{z=0} + \cdots\right) -$$

$$- \frac{1}{n_1^2}\left(\alpha + \frac{(\nabla H)^2}{2} - \cdots\right)\left(\frac{\partial E_1}{\partial z}|_{z=0} + H\frac{\partial^2 E_1}{\partial z^2}|_{z=0} + \frac{H^2}{2}\frac{\partial^3 E_1}{\partial z^3}|_{z=0} + \cdots\right) =$$

$$= \alpha \frac{1}{n_1^2} \left(-\frac{\partial E_1}{\partial z} |_{z=0} + \frac{\partial E_1}{\partial x} \frac{\partial H}{\partial x} |_{z=0} + \frac{\partial E_1}{\partial y} \frac{\partial H}{\partial y} |_{z=0} \right) +$$

$$+ \alpha \frac{1}{n_1^2} \left(-H \frac{\partial^2 E_1}{\partial z^2} |_{z=0} + H \frac{\partial^2 E_1}{\partial x \partial z} |_{z=0} \cdot \frac{\partial H}{\partial x} + \frac{\partial^2 E_1}{\partial y \partial z} |_{z=0} \cdot \frac{\partial H}{\partial y} \right) -$$

$$- \alpha \frac{1}{n_1^2} \left(\frac{\partial E_1}{\partial z} |_{z=0} \frac{(\nabla H)^2}{2} + \frac{H^2}{2} \frac{\partial^3 E_1}{\partial z^3} |_{z=0} + \cdots \right).$$

$$\frac{\partial E_2}{\partial \mathbf{n}} |_{z=H(x,y)} =$$

$$= \alpha \frac{1}{n_2^2} \left(-\frac{\partial E_2}{\partial z} |_{z=0} + \frac{\partial E_2}{\partial x} \frac{\partial H}{\partial x} |_{z=0} + \frac{\partial E_2}{\partial y} \frac{\partial H}{\partial y} |_{z=0} - H \frac{\partial^2 E_2}{\partial z^2} |_{z=0} + H \frac{\partial^2 E_2}{\partial x \partial z} |_{z=0} \cdot \frac{\partial H}{\partial x} \right) +$$

$$+ \alpha \frac{1}{n_2^2} \left(\frac{\partial^2 E_2}{\partial y \partial z} |_{z=0} \cdot \frac{\partial H}{\partial y} - \frac{\partial E_2}{\partial z} |_{z=0} \frac{(\nabla H)^2}{2} + \frac{H^2}{2} \frac{\partial^3 E_2}{\partial z^3} |_{z=0} + \cdots \right). \quad (10.7)$$

Then, a boundary condition of type (10.4) taking into account (10.6) and (10.7) assumes the form

$$\frac{1}{n_1^2} \left(\frac{\partial E_1}{\partial z} |_{z=0} - \frac{\partial E_1}{\partial x} \frac{\partial H}{\partial x} |_{z=0} - \frac{\partial E_1}{\partial y} \frac{\partial H}{\partial y} |_{z=0} + H \frac{\partial^2 E_1}{\partial z^2} |_{z=0} - H \frac{\partial^2 E_1}{\partial x \partial z} |_{z=0} \cdot \frac{\partial H}{\partial x} \right) -$$

$$- \frac{1}{n_1^2} \left(\frac{\partial^2 E_1}{\partial y \partial z} |_{z=0} \cdot \frac{\partial H}{\partial y} + \frac{\partial E_1}{\partial z} |_{z=0} \frac{(\nabla H)^2}{2} + \frac{H^2}{2} \frac{\partial^3 E_1}{\partial z^3} |_{z=0} \right) = \quad (10.8)$$

$$= \frac{1}{n_2^2} \left(\frac{\partial E_2}{\partial z} |_{z=0} - \frac{\partial E_2}{\partial x} \frac{\partial H}{\partial x} |_{z=0} - \frac{\partial E_2}{\partial y} \frac{\partial H}{\partial y} |_{z=0} + H \frac{\partial^2 E_2}{\partial z^2} |_{z=0} - H \frac{\partial^2 E_2}{\partial x \partial z} |_{z=0} \cdot \frac{\partial H}{\partial x} \right) -$$

$$- \frac{1}{n_2^2} \left(\frac{\partial^2 E_2}{\partial y \partial z} |_{z=0} \cdot \frac{\partial H}{\partial y} + \frac{\partial E_2}{\partial z} |_{z=0} \frac{(\nabla H)^2}{2} + \frac{H^2}{2} \frac{\partial^3 E_2}{\partial z^3} |_{z=0} \right).$$

We will seek the reflected field in medium 1 and the field transmitted to medium 2 in the form

$$E_1(x, y, z) = E_{inc}(x, y, z) + \sum_{n=0}^{\infty} E_{01}^n(x, y, z), \quad (10.9)$$

$$E_2(x, y, z) = \sum_{n=0}^{\infty} E_{02}^n(x, y, z), \quad (10.10)$$

(we omit factor $\exp(-i\omega t$ for brevity), where $E_{inc}(x, y, z)$ is the primary monochromatic field incident on the rough surface, $E_{01}^0(x, y, z)$ is the amplitude of the reflected wave, and $E_{02}^0(x, y, z)$ is the amplitude of the transmitted wave. The remaining terms of series (10.9) and (10.10) are propagating and attenuating scattered modes in the upper and lower media. Substituting relations (10.9) and (10.10) into (10.7) and (10.8), we obtain the boundary conditions for successive approximations of the field:

$$E_{inc}|_{z=0} + E_{01}^0|_{z=0} = E_{02}^0|_{z=0}, \tag{10.11}$$

$$E_{01}^1|_{z=0} + H\left(\frac{\partial E_{inc}}{\partial z}\right)_{z=0} + H\left(\frac{\partial E_{01}^0}{\partial z}\right)_{z=0} =$$

$$= H\left(\frac{\partial E_{02}^0}{\partial z}\right)_{z=0} + E_{02}^1|_{z=0}, \tag{10.12}$$

$$E_{01}^2|_{z=0} + H\frac{\partial E_{01}^1}{\partial z}|_{z=0} + \frac{H^2}{2}\left(\frac{\partial E_{inc}^2}{\partial z^2} + \frac{\partial^2 E_{01}^0}{\partial z^2}\right)_{z=0} =$$

$$= E_{02}^0 + H\left(\frac{\partial E_{02}^1}{\partial z}\right)_{z=0} + E_{02}^1|_{z=0} + \left(\frac{H^2}{2}\frac{\partial^2 E_{02}^0}{\partial z^2}\right)_{z=0}, \tag{10.13}$$

$$\frac{1}{n_1^2}\left(\frac{\partial E_{inc}}{\partial z} + \frac{\partial E_{01}^0}{\partial z}\right)_{z=0} = \frac{1}{n_2^2}\left(\frac{\partial E_{02}^0}{\partial z}\right)_{z=0}, \tag{10.14}$$

$$\frac{1}{n_1^2}\left(\left(\frac{\partial E_{01}^1}{\partial z}\right)_{z=0} - \left(\frac{\partial E_{inc}}{\partial x} + \frac{\partial E_{01}^0}{\partial x}\right)_{z=0}\frac{\partial H}{\partial x} - \left(\frac{\partial E_{inc}}{\partial y} + \frac{\partial E_{01}^0}{\partial y}\right)_{z=0}\frac{\partial H}{\partial y}\right) +$$

$$+ \frac{1}{n_1^2}H\left(\frac{\partial^2 E_{01}^0}{\partial z^2}\right)_{z=0} =$$

$$= \frac{1}{n_2^2}\left(\left(\frac{\partial E_{02}^1}{\partial z}\right)_{z=0} - \left(\frac{\partial E_{02}^0}{\partial x}\right)_{z=0}\frac{\partial H}{\partial x} - \left(\frac{\partial E_{02}^0}{\partial y}\right)_{z=0}\frac{\partial H}{\partial y}\right) +$$

$$+ \frac{1}{n_2^2}\left(H\left(\frac{\partial^2 E_{02}^0}{\partial z^2}\right)_{z=0}\right), \tag{10.15}$$

$$\frac{1}{n_1^2}\left(\left(\frac{\partial E_{01}^2}{\partial z}\right)_{z=0} - \left(\frac{\partial E_{inc}}{\partial x} + \frac{\partial E_{01}^1}{\partial x}\right)_{z=0}\frac{\partial H}{\partial x} - \left(\frac{\partial E_{inc}}{\partial y} + \frac{\partial E_{01}^1}{\partial y}\right)_{z=0}\frac{\partial H}{\partial y}\right) +$$

$$+ \frac{1}{n_1^2}H\left(\left(\frac{\partial^2 E_{inc}}{\partial z^2} + \frac{\partial^2 E_{01}^1}{\partial z^2}\right)_{z=0}\right) +$$

$$
+\frac{1}{n_1^2}\left(H\left(\frac{\partial^2 E_{inc}}{\partial x \partial z} + \frac{\partial^2 E_{01}^1}{\partial x \partial z}\right)_{z=0} \cdot \frac{\partial H}{\partial x} - H\left(\frac{\partial^2 E_{inc}}{\partial y \partial z} + \frac{\partial^2 E_{01}^1}{\partial y \partial z}\right)_{z=0} \cdot \frac{\partial H}{\partial y}\right) +
$$

$$
+\frac{1}{n_1^2}\left(\left(\frac{\partial E_{inc}}{\partial z} + \frac{\partial E_{01}^1}{\partial z}\right)_{z=0} \frac{(\nabla H)^2}{2} + \frac{H^2}{2}\left(\frac{\partial^3 E_{inc}}{\partial z^3} + \frac{\partial^3 E_{01}^1}{\partial z^3}\right)_{z=0}\right) =
$$

$$
= \frac{1}{n_2^2}\left(\left(\frac{\partial E_{02}^2}{\partial z}\right)_{z=0} - \left(\frac{\partial E_{02}^1}{\partial x}\right)_{z=0} \frac{\partial H}{\partial x} - \left(\frac{\partial E_{02}^1}{\partial y}\right)_{z=0} \frac{\partial H}{\partial y} + H\left(\frac{\partial^2 E_{02}^1}{\partial z^2}\right)_{z=0}\right) +
$$

$$
\frac{1}{n_2^2}\left(H\left(\frac{\partial^2 E_{02}^1}{\partial x \partial z}\right)_{z=0} \cdot \frac{\partial H}{\partial x} - H\left(\frac{\partial^2 E_{02}^1}{\partial y \partial z}\right)_{z=0} \cdot \frac{\partial H}{\partial y} + \left(\frac{\partial E_{01}^1}{\partial z}\right)_{z=0} \frac{(\nabla H)^2}{2}\right) +
$$

$$
+\frac{1}{n_2^2}\frac{H^2}{2}\left(\frac{\partial^3 E_{01}^1}{\partial z^3}\right)_{z=0}. \tag{10.16}
$$

We assume that the perturbation of the interface between two media is described by a certain periodic function $H(x + 2a, y + 2a) = H(x, y)$. In accordance with the periodicity conditions, function $H(x, y)$ can be expanded into a Fourier series. We assume that the number of harmonics in this series is finite; this gives

$$
H(x, y) = \sum_{m=0}^{M} \sum_{n=0}^{N} H_{mn} \exp(i\lambda_n x) \exp(i\lambda_m y), \tag{10.17}
$$

where

$$
\lambda_n = \frac{\pi n}{a}, \lambda_m = \frac{\pi m}{a}.
$$

Taking into account relations (10.9) and (10.17), we will seek the field in the upper medium in the form

$$
E_1 = \exp(i\tau_{inc}(x, y, z)) + \sum_{m=0}^{M} \sum_{n=0}^{N} B_{mn}^{-} H_{mn} \exp(i\lambda_n x) \exp(i\lambda_m y) \times
$$

$$
\times \exp(i\tau_{ref}(x, y, z)) \tag{10.18}
$$

and the field in the lower medium will be sought, taking into account relations (10.10) and (10.17), in the form

$$
E_2 = \sum_{m=0}^{M} \sum_{n=0}^{N} B_{mn}^{+} H_{mn} \exp(i\lambda_n x) \exp(i\lambda_m y) \exp(i\tau_{tr}(x, y, z)), \tag{10.19}
$$

where

$$\tau_{inc}(x, y, z) = k_{1x}x + k_{1y}y - k_{1z}z, \ \tau_{ref} = k_{1x}x + k_{1y}y + k_{1z}z, \ \tau_{2tr} = k_{2x}x + k_{2y}y - k'_{2z}z,$$

$$k_{1x} = kn_1 \sin(\theta) \sin(\phi), k_{1y} = kn_1 \sin(\theta) \cos(\phi), k_{1z} = kn_1 \cos(\theta),$$

$$k_{2x} = kn_2 \sin(\theta) \sin(\phi), k_{2y} = kn_2 \sin(\theta) \cos(\phi), k_{2z} = kn_2 \cos(\theta).$$

Substituting relations (10.18) and (10.19) into (10.1)–(10.2), we find that these equations hold under the following conditions:

$$k_{1x}^2 + k_{1y}^2 + k_{1z}^2 + \lambda_n^2 + \lambda_m^2 = k^2 n_1^2, \quad k_{2x}^2 + k_{2y}^2 + k_{2z}^2 + \lambda_n^2 + \lambda_m^2 = k^2 n_2^2.$$

We substitute expressions (10.18) and (10.19) into (10.11)–(10.16), multiply the result by (10.11)–(10.16) $\exp(-i\lambda_{n_1} x) \exp(-i\lambda_{m_1} y)$ and integrate over the period; this gives a system of linear equations in B^+ and B^-. Solving the resultant system, we obtain corrections to the amplitude transmission and reflections coefficients, which have the form

$$B^- = B_{00}^- + H^2 B_{00}^- = \left(1 + \sum_{n=0}^{N} \sum_{m=0}^{M} H_{mn}^2 k_1 (-2k_2 + 2\alpha_{ref} - \alpha_{tr})\right) B_{00}^- \quad (10.20)$$

$$B^+ = B^+ + H^2 B_{00}^- = \left(1 + 0.5 k_1 k_2 \sum_{n=0}^{N} \sum_{m=0}^{M} H_{mn}^2 ((k_1 - k_2) + 2\alpha_{ref} - 2\alpha_{tr})\right) B_{00}^+,$$
$$(10.21)$$

where $k_1 = kn_1, k_2 = kn_2, \alpha_{ref} = k_{1x} + k_{1y} + k_{1z}, \alpha_{tr} = k_{2x} + k_{2y} - k'_{2z}$.
Substituting σ^2 for H_{mn}^2 in expressions (10.20)–(10.21), we obtain

$$B^- = B_{00}^- + H^2 B_{00}^- = (1 + \sigma^2 k_1 (-2k_2 + 2\alpha_{ref} - \alpha_{tr})) B_{00}^-, \quad\quad (10.22)$$

$$B^+ = B_{00}^+ + H^2 B_{00}^- = (1 + 0.5 k_1 k_2 \sigma^2 ((k_1 - k_2) + 2\alpha_{ref} - 2\alpha_{tr})) B_{00}^+, \quad (10.23)$$

where B^+ is the amplitude of the reflected wave for the rough interface between the two media and B^- is the amplitude of the transmitted wave for the rough interface between the media. We define σ as the standard deviation of the rough interface profile from the unperturbed boundary.

Having determined the corrections to the amplitude transmission and reflection coefficients, we formulate the problem of reflection of a plane wave from a layer with a slowly varying thickness taking into account the roughness of the surface.

Let us consider an optical system. The system consists of two regions with different refraction indices. To attain the maximal conformity with the structure of the actual object of investigation, we represent the interface between the layer of the model medium in the form of a undulated surface $z = H(x, y)$, where

$H(x, y) = c \sin(ax + by)$, a, b and c are certain arbitrarily defined constant, such that $a \ll 1, b \ll 1, c \ll 1$.

Let us suppose that a plane s- or p- polarized wave is incident on the layer at an angle θ. We consider only the case of the p polarization. We must find the reflected field. We will seek the reflected field in the form of waves with slowly varying amplitudes and rapidly oscillating phases:

$$E_1 = \exp\left(\frac{i}{\varepsilon}\tau_{inc}(\xi_1, \xi_2, \xi_3)\right) + \exp\left(\frac{i}{\varepsilon}\tau_{1ref}(\xi_1, \xi_2, \xi_3)\right) \times$$

$$\times A(\xi_1, \xi_2, \xi_3, \varepsilon_x, \varepsilon_y), \tag{10.24}$$

$$E_2 = \exp\left(\frac{i}{\varepsilon}\tau_{2tr}(\xi_1, \xi_2, \xi_3)\right) B^+(\xi_1, \xi_2, \xi_3, \varepsilon_x, \varepsilon_y) +$$

$$\exp\left(\frac{i}{\varepsilon}\tau_{3ref}(\xi_1, \xi_2, \xi_3)\right) B^-(\xi_1, \xi_2, \xi_3, \varepsilon_x, \varepsilon_y), \tag{10.25}$$

$$E_3 = \exp\left(\frac{i}{\varepsilon}\tau_{3tr}(\xi_1, \xi_2, \xi_3)\right) C(\xi_1, \xi_2, \xi_3, \varepsilon_x, \varepsilon_y). \tag{10.26}$$

We seek amplitudes A and C in the form of power series in small parameters ε_x, ε_y (see Chap. 4) It should be noted that the expressions for amplitudes $B^\pm$ taking into account relations (10.22)–(10.23) have the form

$$B^+(\xi_1, \xi_2, \xi_3, \varepsilon_x, \varepsilon_y) = \sum_{i=0}^{\infty}\sum_{j=0}^{\infty} B^+_{(00)ij}(\xi_1, \xi_2, \xi_3) \times$$

$$\times (1 + F_1)(\varepsilon_x^i \cdot \varepsilon_y^j), \tag{10.27}$$

$$B^-(\xi_1, \xi_2, \xi_3, \varepsilon_x, \varepsilon_y) = \sum_{i=0}^{\infty}\sum_{j=0}^{\infty} B^-_{(00)ij}(\xi_1, \xi_2, \xi_3) \times$$

$$\times (1 + F_2)(\varepsilon_x^i \cdot \varepsilon_y^j), \tag{10.28}$$

where $F_1 = 0.5k_1k_2\sigma^2((k_1 - k_2) - 2\alpha_{ref} + 2\alpha_{tr})$, $F_2 = \sigma^2 k_1(-2k_2 - 2\alpha_{ref} + \alpha_{tr})$.

Note that the expressions for amplitudes A, C, $B^\pm$ are defined analogously to the method described on Chap. 4. Substitution of expressions (10.24)–(10.26) into (4.6)–(4.11) generates a recurrent system of equations. From this system for the reflected field, we find reflection coefficient A taking into account the roughness of the interface with the medium being simulated.

The expression for the reflection of a Gaussian beam with an arbitrary cross section is defined analogously to the method described in Chap. 4. We note that the expression which connects the frequency of natural oscillations of the optical resonator loaded with the sample of the biological tissue under investigation with electrophysical parameters of this biological structure such as real and imaginary parts of their refractive indices and sizes are described in Chap. 9.

10.3 Numerical Calculations for a Resonator with Chosen Parameters and Conclusions

Let us consider an optical resonator with a model medium (sample of biotissue) with the following parameters: the distance $L = 11$ cm between the mirrors, radii of the mirrors are $M_1 = 100$ cm and $M_2 = 46.3$ cm. The arbitrarily chosen constants are $a = -0.0024$, $b = 0.020$, $c = 10^{-2}$. The values of parameters a, b and c are chosen for the interface between the layer being simulated so that the shape of the surface is in the best conformity with the shape of the interface of the corresponding layer in the structure of the biological sample being simulated; the thickness of the sample being simulated was $0.3\,\mu$m. All calculations were made for the fundamental transverse mode of a linear resonator.

Figure 10.1a, b show the dependence of the absorption coefficient of the biological sample being simulated on the wavelength for $\sigma = 0$ and $\sigma = 0.3$ nm, where σ is defined as the standard deviation of the profile of the rough boundary from the unperturbed boundary. It follows from the graphs that the absorption coefficient of

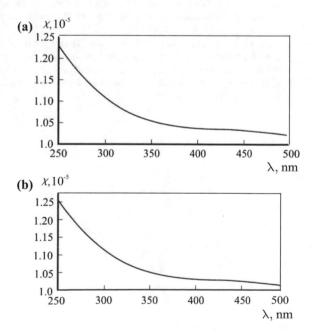

Fig. 10.1 Dependence of the absorption coefficient of the biological sample being simulated on the wavelength for the following parameters of the model medium: the real part of the refractive index of the sample being simulated is 1.3, $\sigma = 0$ (**a**), $\sigma = 0.3$ nm (**b**)

the biological sample in the ultraviolet range is large, while the absorption coefficient in the visible range decreases and remains almost unchanged.

Thus, the model constructed here makes it possible to analyze the biophysical characteristics associated with absorption of light in optically thin layers on account of small-scale inhomogeneities. This also makes it possible to vary (on the same setup) the biological objects and their electrophysical parameters, as well as characteristic thicknesses of the layers and the characteristic sizes of roughnesses of the biological structure to determine the dependence between these parameters. Using this approach systematically, it will probably be possible to find correlations between electrophysical parameters of the biological structure being simulated and its biological properties.

It should be noted that by varying the absorption coefficient of the biological tissue, one can use this model for in vitro measurements of the spectral characteristic of the biological tissue taking into account small-scale inhomogeneities to construct the spectral autograph for determining pathological changes in the biological samples under investigation.

Analogous dependences can be calculated for lasers with other parameters and used for processing of experimental absorption curves for biological structures under investigation taking into account small-scale inhomogeneities.

References

1. K.G. Kulikov, Control of the optical characteristics of an optically thin layer with a rough surface by intracavity laser spectroscopy, in *European Conferences on Biomedical Optics*, 22–26 May 2011, Munich, Germany (2011)
2. K.G. Kulikov, Accounting for small-scale inhomogeneities in the simulation of electrophysical characteristics of an optically thin layer method intracavity laser spectroscopy, in *SPIE Photonics Europe*, 16–19 April 2012 Brussels, Belgium (2012)
3. S.M. Rytov, Yu.A. Kravtsov, V.I. Tatarskii, Introduction to Statistical Radiophysics. Part II. Moscow (1978), 463 p

Chapter 11
Simulation of the Thermal Processes

Abstract We propose the mathematical model for calculation of the hyperthymia of a multilayer biological structure under the action of laser radiation. For the case in vivo, the dependences of the temperature field on the refractive index and absorption coefficient of the biological tissue under study (epidermis, the upper derma layer, the lower derma layer, blood and its corpuscles) are determined. The obtained quantitative estimates can be used to predict the changes in the optical properties of the biological structure that are caused by the biophysical, biochemical, and physiological processes during the action of a nonpolarized monochromatic radiation flow on the structure surface.

11.1 Introduction

Laser therapy belongs to promising and dynamically developing fields in modern medicine. The therapeutic action of laser radiation is related to the hypothermia of biological tissue, which requires a model for the calculation of the temperature field in the tissue subjected to low-intensity (noncoagulating) laser radiation. There exist a number of works dealing with the problems of the mathematical simulation of the laser radiation distribution in multilayer biological tissue and with the related thermal processes. In most works [1, 2], researchers have calculated the temperature fields appearing during the irradiation of biological tissue by a low-intensity laser beam at various times. For example, to find the depth profile of the absorbed energy in irradiated tissue, researchers used various numerical methods, including the discrete coordinate method [3], finite-difference schemes [1], the Green function method [1], and the Monte Carlo method [4]. The last method is effective for complex geometry of a biological sample.

However, an analysis of the thermal effect of laser radiation should not be purely physical, since it has to include biological (biophysical) studies of the response of the organism. Thus, the problem of the thermal effect of laser radiation can be divided into four problems to be solved successively [5]:

(1) the description of the laser radiation energy distribution;
(2) the determination of the absorption characteristics of the biological material;
(3) an analysis of the temperature distribution in irradiated tissue; and
(4) the study of the biological (biochemical, physiological) changes in the tissue caused by an increase in the temperature.

In this chapter we construct a mathematical model that can vary the electrophysical parameters of a biological structure (the real and imaginary parts of the refractive indices of blood and its corpuscles, epidermis, the upper derma layer, the lower derma layer) and the characteristic sizes of blood corpuscles and can find relations between them and the biological properties of blood by allowing for the laser-induced heating of biological tissue. As a result, we can perform in vivo analysis of the temperature distribution as a function of the electrophysical parameters of the biological structure under study. In the first part of this work, we consider the problem of the scattering of a plane electromagnetic wave by a three-layer spherical particle simulating a blood cell (see Chap. 3). In the second part, we analyze the more complex case of the reflection of a plane wave using a biological sample consisting of two continuous layers and one layer with heterogeneous inclusions that simulate blood cells with different refractive indices and briefly examine the problem of the reflection of a Gaussian beam with an arbitrary cross section under the conditions given above and the problem of determining the dependence of the radiation intensity on the refractive index for a system of blood vessels located in the upper derma layer (see Chaps. 4, 6). These parts have an auxiliary character. In the third part, we solve the problem of the heating of a blood vessel under the action of a laser beam incident on the outer surface of a biological structure.

Chapter is based on the results of the [6, 7].

11.2 Mathematical Model for Heating of Biological Tissue by Laser Radiation

We propose a mathematical model for the heating of a blood vessel by laser radiation incident on the outer skin surface. In this model, we use dimensional variables. The laser radiation incident on the skin surface is absorbed by the biological tissue layers (epidermis, derma) and the blood hemoglobin, increasing the temperature in the subskin layers and inside blood vessels. In the general case, the simulation of the thermal processes in biological tissue requires the solution of the three-dimensional equation

$$(c \cdot \rho)^{-1} \cdot \mathrm{div}(\lambda \cdot \mathrm{grad} T(\mathbf{r}, t)) + Q(\mathbf{r}, m_\tau^j, x_\tau^j) = \frac{\partial T}{\partial t}, \qquad (11.1)$$

where c is the specific heat, ρ is the density, λ is the thermal conductivity, $m_\tau^j = N_\tau^j / n_o$, N_τ^j is the complex refractive index of the jth particle for the τth concentric layer, n_o is the refractive index of the environment, $x_\tau^j = ka_\tau^j, j = \overline{1...N}$, $\tau = \overline{1,3}$, where a_τ^j is the radius of the jth particle with τth concentric layers, $T(\mathbf{r}, t)$ is the desired temperature distribution and $Q(\mathbf{r}, m_\tau^j, x_\tau^j)$ is the volume power density distribution of the heat loads in the biological tissue that are induced by its absorption. This distribution was found at the stage of solving the optical problem. We write $Q(\mathbf{r}, m_\tau^j, x_\tau^j)$ in the form [8]:

$$Q(\mathbf{r}, m_\tau^j, x_\tau^j) = \frac{E_0}{\tau_p} \mu \phi(\mathbf{r}, m_\tau^j, x_\tau^j),$$

where $\phi(\mathbf{r}, m_\tau^j, x_\tau^j)$ is a intensity at $\mathbf{r}(x, y, z)$, divided to a unit power density on the surface in the simulated biological structure, $\int_{4\pi} I(\lambda, m_\tau^j, x_\tau^j, \mathbf{r}, \theta, \varphi) d\Omega$ is intensity, where $I(\lambda, m_\tau^j, x_\tau^j, \mathbf{r}, \theta, \varphi)$ determined from (4.69), $d\Omega = \sin\theta d\theta d\varphi$ is the solid angle, μ is the absorption coefficient of the medium, E_o is the radiation power density and τ_p is pulse duration.

Since the model includes a few of the skin layers, then (11.1) has been solved for each of them separately. For areas where vessels are not anatomically, thermal calculation was based on (11.1) . In areas in which vessels are present (in the upper layer of the dermis) we added more heat sources, which are caused by the flow of blood. In this layer, we have

$$(c \cdot \rho)^{-1} \cdot \mathrm{div}(\lambda \cdot \mathrm{grad}T(\mathbf{r}, t)) + Q(\mathbf{r}, m_\iota^j, x_\iota^j) + Q_{blood}(\mathbf{r}, t, T) = \frac{\partial T}{\partial t}, \quad (11.2)$$

where

$$Q_{blood}(\mathbf{r}, t, T) = c \cdot \rho(\rho_{blood} f(t, T) \cdot (T_{blood} - T(\mathbf{r}, t)),$$

ρ_{blood} is density blood, T_{blood} is temperature blood, $f(t, T)$ is the density of the flow of blood into the tissues.

Let's write the boundary conditions. In a linearized form, the interaction of the outer skin surface with the environment (convection) can be described by the boundary conditions of the third kind [1]

$$\left(\lambda \frac{\partial T}{\partial z} - A(T - T_o)\right)|_{z=0} = 0, \quad (11.3)$$

where A is the reduced heat-transfer coefficient and, T_0 is the initial temperature.

$$T|_{t=0, z=0} = 34°, \quad T|_{t=0, z=h_1(x,y)} = 37°. \quad (11.4)$$

Expression (11.4) means that the temperature changes with the depth from 34° to 37°.

At the interface between the ith and $(i + 1)$th layers $(z = h_i(x, y))$, the following continuity conditions of the heat flow and temperature are met:

$$\left(\lambda_i \frac{\partial T_i}{\partial z} - \lambda_{(i+1)} \frac{\partial T_{(i+1)}}{\partial z}\right)|_{z=h_i(x,y)} = 0 \tag{11.5}$$

$$\left(T_i - T_{(i+1)}\right)|_{z=h_i(x,y)} = 0, \tag{11.6}$$

where $h_i(x, y)$ is determined from (4.1).

When solving set of (11.1)–(11.6), we obtain

(a) the temperature distribution over layers along the propagation direction of the laser beam; and
(b) the dependence of the temperature on the optical properties of the biological tissue, which can be used to study the effect of the temperature field on the electrophysical parameters of the biological tissue for the case in vivo.

For further investigation and analysis of the dependences obtained, we will use numerical methods.

11.3 Numerical Calculations Using a Model Medium and Conclusions

To numerically solve the set of (11.1)–(11.6), we construct an implicit iteration scheme on a spacetime mesh, the boundary conditions for temperature being replaced by their finiteanalogs [9]. We consider the model medium that is shown in Fig. 6.1 and has the following parameters [10]: the characteristic layer thicknesses are $d_2 = 65 \cdot 10^{-6}$, $n_2^\circ = 1.50$, $n_3^\circ = 1.40$, $n_4^\circ = 1.35$, $n_5^\circ = 1.40$ $n_1^\circ = 1$, $\chi_1 = 0$, $\chi_2 = \chi_3 = \chi_4 = \chi_5 = 10^{-5}$, the wavelength is $\lambda = 0.63$ μm (center of the line of a He−Ne laser). The arbitrarily specified constants are $a_1 = -0.0024$, $b_1 = 0.020$, $a_2 = 0.021$, $b_2 = 0.030$, $a_3 = 0.041$, $b_3 = 0.051$, $c_1 = c_2 = c_3 = 10^{-2}$. The values of parameters a_1, b_1, a_2, b_2, a_3, b_3, c_1, c_2 and c_3 are chosen for the interface of each layer so that the surface shape are as close as possible to the interface shape of the corresponding layer in the structure of human skin, the thermal conductivity (W/(m K)), the specific heat J/(kg), and the density $\times 10^{-3}$(kg/m^3) are 0.498, 3.2 and 1 for the first layer, 0.266, 3.7 and 1.6 for the second layer, 0.530, 3.6 and 1 for the third layer, 0.266, 3.7 and 1.6 for the fourth layer, and the heat-transfer coefficient is 0.009 W cm^{-2} K^{-1}. The calculations were performed for two-layer particles simulating red corpuscles. Each layer was taken to have ten particles, the speed of blood flow in the dermis is 15 mL/(min 100 g), pulse duration is 20c, the radiation power density is 1W/m^2.

Figure 11.1a and b shows the time-dependent temperature distribution in the direction of the incident radiation (z axis) for a multilayer light-absorbing and scattering medium that simulates human skin and its components at various refractive indices. The upper layer of the simulated biological tissue (epidermis) is seen to be

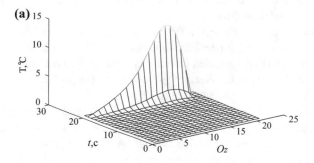

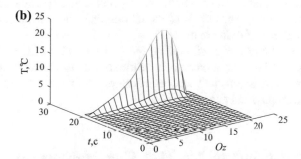

Fig. 11.1 Spatial temperature distribution in the incident radiation direction as a function of time t in the vicinity of the heliumneon laser radiation line (0.63 μm). At the initial time $t = 0$, the temperature of the medium is 34°. The medium parameters are as follows: the real value of the refractive index of the cytoplasm of the biological particle is **a** 1.35 or **b** 1.45, the imaginary value of the refractive index of the refractive index of the cytoplasm of the biological particle is 0.0001, the real value of the refractive index of the plasma membrane of the biological particle is **a** 1.23 or **b** 1.43, the imaginary value of the refractive index of the plasma membrane of the biological particle is 0.0001, the radius of the cytoplasm of the biological particle is 3 μm, and the radius of the plasma membrane of the biological particle is **a** 2 or **b** 2.3 μm. **b** The real value of the refractive index of the plasma membrane of the biological particle is 1.43, the imaginary value of the refractive index of the plasma membrane of the biological particle is 0.0001, the radius of the cytoplasm of the biological particle is 3 μm, and the radius of the plasma membrane of the biological particle is 2.3 μm

significantly heated, which is likely to be related to the fact that the light is strongly absorbed by melanin in the surface layer at the given wavelength (see Chap. 1). We can also conclude that the surface temperature exceeds approximately 45° at the tenth second of the continuous action of laser radiation (Fig. 10.1b), and tissue necrosis or thermal burn can appear. Thus, with the model, we can estimate the thermal action of laser radiation on biological tissue, choose the optimum action time to provide uniform and long-term heating of the tissue by excluding negative reactions, and determine the boundaries of destruction and tissue necrosis. It should be noted that our mathematical model is rather sensitive to the changes in the refractive indices of the simulated biological tissue and its components that are induced by nonpolarized monochromatic radiation flow.

The following effect obtained in the model experiment is of interest. We got a linear relation between the temperature field distribution of the incident radiation and the refractive indices of the simulated biological tissue. As a result, this model can be used to predict changes in the electrophysical properties of the biological structure subjected to laser radiation for the case in vivo.

Our model can vary the composition, the electrophysical parameters, the thermo-physical characteristics, and the characteristic layer thickness of biological objects, as well as the characteristic sizes of the biological structure under study, in one apparatus in order to analyze the biophysical processes related to the thermal action of laser radiation on the upper skin layers.

Using such a simulation, we can both find the preliminary parameters of the laser radiation field and reveal the effects of the responses to laser irradiation at various levels of organization of living matter.

On the whole, the results of simulating the thermal fields of laser radiation can be used to improve laser thermotherapy and biostimulation methods and can serve as the basis for the mathematical support of the experimental determination of the optical and thermophysical parameters.

11.4 The Mathematical Model of Thermo-chemical Denaturation of Biological Structure

The results of calculations of the temperature field in a simulated biological structure can be used to assess the kinetics of denaturation of tissue. Note that the models of thermo-chemical denaturation of biological structures such as corneal tissue and skin was considered in [11, 12]. The correct solution to estimate the kinetics of thermal decomposition of biological structures is difficult, because the biochemical composition of the cells is complex. However, the necessary practical estimates accuracy could be achieved with the introduction of a number of assumptions [13]. The basis of biochemical reactions stimulated by heat, are such processes as break chemical bonds, the conformational transition. This class includes reactions and thermal denaturation of proteins and lipids, enzymes, etc.

To describe such reactions we use the kinetic equation of irreversible chemical reaction of the first order, where the temperature dependence of the reaction rate constant $K(T)$ is Arrhenius law:

$$\frac{df}{dt} = -K(T)f, \quad K(T) = \frac{kT}{h} \exp\left(\frac{-\Delta H - T\Delta S}{RT}\right), \tag{11.7}$$

Fig. 11.2 The threshold radiant exposure

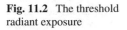

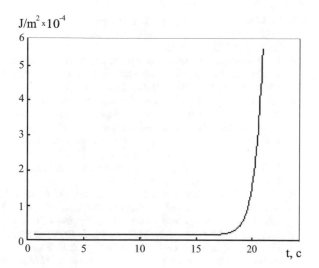

where f is the relative concentration of the protein molecules, t is time, ΔH is enthalpy of activation, ΔS is entropy of activation, R is universal gas constant, h is Planck constant, k is Boltzmann constant.

When solving equations (11.1)–(11.6) and (11.7) it is possible to determine the radiant exposure which causes the primary disorders, simulated biological structure, in particular, the dermis.

Criterion for such a disorder is decrease in the dimensionless concentration of the original protein of the initial value $f = 1$ before $f = \exp(-1)$. This value radiant exposure is the threshold.

Figure 11.2 shows the calculated dependence of the threshold energy density of helium-neon laser on the laser pulse duration, ΔH is 430000 J/mol and ΔS is 940 J/(mol K) [14, 15]. As follows from the figure, with increasing duration of exposure there is a sharp increase in the consumption of energy required for create the threshold conditions of coagulation. This phenomenon can be explained by the loss of selectivity effects, spreading the temperature field, and as a consequence, increasing the heated volume. Thus, the mathematical model can be considered for use the development of optimal regime and technical characteristics of lasers used in biomedical research.

References

1. YuN Scherbakov, A.N. Yakunin, I.V. Yaroslavsky, V.V. Tuchin, Modeling of thermal processes in the interaction of laser radiation with noncoagulating multilayer biological tissue. part 1. Opt. Spectrosc. **76**(5), 845–850 (1994)
2. A.Yu. Seteykin, I.V. Krasnikov, Calculation of temperature fields arising from the interaction of laser radiation with multilayer biological material. J. Opt. Technol. **73**(3), 31–34 (2006)

3. M. Motamedi, S. Rastegar, G.L. Le Carpentier, A.J. Welch, Light and temperature distribution in laser irradiated tissue: the influence of anisotropic scattering and refractive index. Appl. Opt. **28**(2), 2230–2237 (1989)
4. I.V. Meglinski, Simulation of the reflectance spectra of optical radiation from a randomly inhomogeneous multilayer strongly scattering and absorbing light environments using the Monte Carlo. Quantum Electron. **31**(12), 1101–1107 (2001)
5. S.D Pletnev, *Lasers in Clinical Medicine* (Moscow, 1996), p. 427
6. K.G. Kulikov, The modeling of the temperature field, formed inside multilayer biological tissue under the affect of the laser emission, in *Proceedings of SPIE*, vol. 7373 (2009)
7. K.G. Kulikov, Simulation of the thermal processes induced by action of laser radiation on organic media. Tech. Phys. **54**(2), 259–267 (2009)
8. M.Z. Smirnov, A.E. Pushkareva, The influence of blood flow to the laser heating of the skin. Opt. Spectrosc. **99**(5), 875–878 (2005)
9. A.A. Samarskii, *Theory of Difference Schemes* (Moscow, 1989); (Dekker, New York, 2001)
10. V.V. Tuchin, *Lasers and Fiber Optics in Biomedical Studies* (Saratovsky Univ, Saratov, 1998)
11. F.S. Barnes, *Biological Damage Resulting from Thermal Pulses, Laser Applications in Medicine and Biology* ed. by M.L. Wolbarsht (Plenum Press, New York, 1974)
12. G.I. Zheltov, V.N. Glazkov, A.I. Kirkovsky, A.S. Podol'tsev, Mathematical models of laser-tissue interactions for treatment and diagnosis in ophthalmology, in Laser application in life sciences, part two, *Proceedings of SPIE*, vol. 1403 (1990) p. 752
13. A. Podol'tsev, G.I. Zheltov, The impact IR radiation on the cornea of the eye. Quantum Electron. **16**, 2136 (1989)
14. R. Agah, J.A. Pearce, A.J. Welch, M. Motamedi, Rate process model for arterial tissue thermal damage: implications on vessel photocoagulation. Las. Surg. Med. **15**, 176 (1994)
15. G.I. Zheltov, L.G. Astafeva, A. Carsten, Laser blocking blood flow: the physical model. Opt. Spectrosc. **102**(3), 518–523 (2007)

Chapter 12
Determination of the Optical Parameters on the Basis of Spectrophotometric Data

Abstract The mathematical model is proposed for determination of the optical parameters on the basis of spectrophotometric data.

12.1 Introduction

Modern medical technologies are based on fundamental research in biophysics, physics, mathematics, chemistry, and biology. The rapid development of new optical methods used in various fields of biology and medicine to study the permeability of cell membranes, the diffusion of substances in cellular structures, the photodynamic and photothermal destruction of cells and tissues, as well as to develop new approaches in photodynamic therapy, optical tomography, optical biopsy etc., drives the need to determine the biophysical characteristics of biological tissues.

Knowledge of the optical characteristics of biological tissues is one of the key factors in the development of mathematical models that adequately describe the propagation of light in biological tissues, which in turn is of fundamental importance for the development of new optical methods used in various fields of biology and medicine. Note that non-invasive spectrophotometry methods allow in vivo (in situ) estimation of the biochemical composition of human soft tissues and their dynamics over time, including the study of short-term and rhythmic fluctuations of all the observed parameters that arise as a result of rhythmic work of the cardiovascular and neuro-reflex systems. The most easily determined parameters in the tissues are: the percentage of different hemoglobin fractions (oxyhemoglobin, reduced hemoglobin, etc.) in the blood, the water saturation of tissues (their hydration), the content of melanin, fat, collagen, keratin, porphyrins and a number of other important enzymes in the surface tissues. The study of short-term fluctuations in the parameters of peripheral microhemodynamics on time intervals of 3–5 min allows us to evaluate the functional state of the vascular bed of biological tissues. And the evaluation of long-term changes in the recorded parameters throughout the day, weeks and months

© Springer International Publishing AG, part of Springer Nature 2018   179
K. Kulikov and T. Koshlan, *Laser Interaction with Heterogeneous Biological Tissue*, Biological and Medical Physics, Biomedical Engineering, https://doi.org/10.1007/978-3-319-94114-1_12

allows monitoring the effectiveness of the patient's treatment and evaluating the effect of various individual treatment procedures. Therefore, non-invasive laser diagnostics (spectrophotometry) in medicine can be effective in a wide variety of fields, from oncology and dermatology to occupational pathology, physiotherapy and other areas of medicine. In clinical medical practice, spectrophotometry is used to diagnose the functional state of biological tissues and organs. The method has such advantages as non-invasiveness and a significant depth of penetration of probing radiation in the red and near-infrared range.

Spectrophotometry as a method is based on a transmission of radiation through the sample under study and recording backscattered radiation. The recorded attenuated radiation contains information about the properties of the bio-object, primarily about the absorption and scattering of radiation in the tissue. In the modern technical implementation, the method makes it possible to quantify the optical parameters (refractive index and absorption coefficient) of biological tissue. Thus, having information on the spectral dependence of these parameters, one can reveal the dynamics of the physiological, morphological and biochemical characteristics of biological tissues. In particular, the analysis of the absorption coefficient spectra of biological tissues makes it possible to determine the concentration of endogenous chromophores (melanin, hemoglobin, bilirubin, etc.).

In this chapter, we solve the following problem: on the basis of spectrophotometric data of reflection $R(\lambda)$ for n measurements intensity of the reflected waves to develop a numerical method (for all the investigated diapason wavelength) for determination n_j(refractive index) jth layer etc. Note that this task is a the inverse problem.

12.2   Algorithm for Solving the Inverse Problem

An algorithm for solving the inverse problem consist of approximation the imaginary part the dielectric constant linear combination of basis functions and the use of the Kramers–Kronig relation for the calculation of the real part this function.

$$\Im\varepsilon_j = \varepsilon_{0_j} + \sum_{i=1}^{n} A_{i_j} \exp[-(\omega - \omega_{i_j})^2/\Delta_{i_j}], \Re\varepsilon_j = 1 + \frac{1}{\pi} v.p. \int_{-\infty}^{+\infty} \frac{\Im\varepsilon_j}{\omega^* - \omega} d\omega^*,$$

where $\varepsilon_{0_j}, \omega_{i_j}, \Delta_{i_j}, A_{i_j}$ are desired parameters, by which optimization is performed.

$$\Theta(\varepsilon_j) = \int_{\lambda} \sum_{i=1}^{n} [R_i(\lambda) - R_i(\lambda, \varepsilon_j)]^2 d\lambda \longrightarrow \min,$$

where $R_i(\lambda, \varepsilon_j)$ is coefficient reflection of the simulated biological tissue (see Chaps. 3–7), ε_j determined from the relations Kramers–Kronig. Thus, it is possible to determine from the measured intensities the reflected waves, the complex refractive index of the jth layer of the simulated biological tissue. Since the absorption coefficient (imaginary part of the refractive index) and the real part of the refractive index can be expressed through real and imaginary parts of the dielectric constant:

$$n_j = \frac{1}{\sqrt{2}}\left[\left[\Re\varepsilon_j^2 + \Im\varepsilon_j^2\right]^{1/2} + \Re\varepsilon_j\right]^{1/2}, \chi_j = \frac{1}{\sqrt{2}}\left[\left[\Re\varepsilon_j^2 + \Im\varepsilon_j^2\right]^{1/2} - \Re\varepsilon_j\right]^{1/2},$$

χ_j is absorption coefficient, n_j is real part refractive index jth layer.

The following is a general structure models interaction of laser radiation with a biotissue for determination coefficient reflection of the simulated biological configuration.

12.3 General Structure Models Interaction of Laser Radiation with a Biotissue

Study of Optical Characteristics of Blood Formed Elements Using Intracavity Laser Spectroscopy for Case In vitro

Input parameters:

$m_1^{j(\lambda)}, m_2^{j}(\lambda)$ are complex refractive index of cytoplasm and nucleus for jth particles,
d is shifted nucleus of jth particle,
ρ is thickness of the layer,
M_1, M_2 ara the radii of mirrors,
L is the mirror distance.

Output parameters:

$\omega = \omega(m_1^{j}(\lambda), m_1^{j}(\lambda), d)$ are frequencies of the resonator eigenmodes

Experimental measurement:

Frequencies of the resonator eigenmodes,
Absorption spectra of the nucleus, cytoplasm, and blood cells.

Calculate:

1. Stokes parameters are highly sensitive not only to the refractive index of the particles with a nonconcentric inclusion but also to the position of the nucleus.
2. $m_1^{j}(\lambda)$ is dependence of the imaginary and of the real parts of the index refraction of the nucleus, cytoplasm of blood cells for different values of d.

An Electrodynamic Model of the Optical Characteristics of Blood and Capillary Blood Flow Rate for Case In vivo

Input parameters:

$n_j(\lambda) + i\chi_j(\lambda)$ is refractive index of the jth layer,
d_j is thickness of the jth layer,
λ is the wavelength.

Output parameters:

$I = I(\upsilon_x, t, n_j, \chi_j, d_j)$ are the dependences of the intensity of the laser radiation
on the refractive index and absorption coefficient
for the system of blood vessels in the upper dermis

Experimental measurement:

$I = I(\upsilon_x, t, n_j, \chi_j, d_j)$ is reflected signal on simulated biological structures.

Calculate:

1. $\upsilon_x(t)$ is rate of blood flow in a capillary at the time instant t,
2. $I = I(n_j, \chi_j, d_j)$ are the spectral characteristics for simulated biological structures.

**Study of the Optical Characteristics of a Biotissue with Large-Scale Inhomo-
geneities for Case In vivo**

Input parameters:

$n_j(\lambda) + i\chi_j(\lambda)$ is refractive index of the jth layer,
d_j is thickness of the jth layer,
λ is the wavelength,
D is fractal dimension,
σ is the standard deviation,
q is the parameter of the spatial-frequency scaling,
M, N are the numbers of harmonics.

Output parameters:

$I = I(n_j, \chi_j, d_j, \sigma, D, q)$ is the dependences of the laser radiation intensity
on the refractive index and absorption coefficient
for a system of blood vessels in
for various absorption
coefficients of the epidermis and dermis.

Experimental measurement:

$I = I(n_j, \chi_j, d_j, \sigma, D, q)$ is reflected signal on simulated
biological structures.

Calculate:

$I = I(n_j, \chi_j, d_j, \sigma, D, q)$—the dependences of the laser radiation intensity
on the refractive index and absorption coefficient
for a system of blood vessels in
the outer layer of the dermis for various absorption
coefficients of the epidermis and dermis.

**Light Scattering by Dielectric Bodies of Irregular Shape in a Layered Medium
for Case In vivo**

Input parameters:

$n_j(\lambda) + i\chi_j(\lambda)$ is refractive index of the jth layer,
d_j is thickness of the jth layer,
λ is the wavelength,
r_1^i is the radius of the cell nucleus of the ith particle,
r_2^i is the radius of the plasma membrane of the ith particle,
$m_1^i(\lambda)$ is refractive index of the cell nucleus of the ith particle,
$m_2^i(\lambda)$ is refractive index of the plasma membrane
of the ith particle,
H is the hematocrit in the capillary,
f is the volume concentration of hemoglobin in erythrocytes,
S is the degree of oxygenation of blood.

Output parameters:

$I = I(n_j, \chi_j, m_1^i, m_2^i, r_1^i, r_2^i, S, H, f, d_j)$ is the dependence of the laser radiation
intensity on the refractive index
epidermis, derma, blood corpuscles,
theoretical determination the function of size distribution for blood cells.

Experimental measurement:

$I = I(n_j, \chi_j, m_1^i, m_2^i, r_1^i, r_2^i, S, H, f, d_j)$ is reflected signal on simulated biological
structure

Calculate:

1. $I = I(n_j, \chi_j, m_1^i, m_2^i, r_1^i, r_2^i, S, H, f, d_j)$,
2. K_{HbO_2} is the normalized spectra of action of laser radiation on oxyhemoglobin,
3. K_{Hb} is the normalized spectra of action of laser radiation on deoxyhemoglobin,
4. the function of size distribution for blood cells.

Modeling of the Optical Characteristics Fibrillar Structure for Case In vivo

Input parameters:

$n_j(\lambda) + i\chi_j(\lambda)$ is refractive index of the jth layer,
d_j is thickness of the jth layer,
λ is the wavelength,
a_1^i is radius ith particle (red blood cell),
a_2^i is radius ith cylinder (collagenic fibers)
$m_i(\lambda)$ is refractive index of ith
red blood cell.

Output parameters:

$I = I(n_j, \chi_j, m_i, a_1^i, a_2^i, d_j)-$ is the dependences of the laser radiation intensity
on the refractive index and absorption coefficient
epidermis, derma, blood corpuscles.

Experimental measurement:

$I = I(n_j, \chi_j, m_i, a_1^i, a_2^i, d_j)$ is reflected signal on simulated
biological structures.

Calculate:

$I = I(n_j, \chi_j)$ are spectral measurements on simulated
biological structures for m_i, a_1^i, a_2^i, d_j.

Study of Optical Properties of Biotissues by the Intracavity Laser Spectroscopy Method for Case In vitro

Input parameters:

$n_j(\lambda) + i\chi_j(\lambda)$ is refractive index of the jth layer,
d_j is thickness jth layer,
M_1, M_2 are the radii of mirrors,
L is the mirror distance

Output parameters:

$\omega = \omega(d_j, n_j, i\chi_j)$ are frequencies of the resonator eigenmodes

Experimental measurement:

$\chi_j(\lambda)$ is dependence of the imaginary part of the refractive index
on simulated biological structures for jth layer.

Calculate:

1. $n_j(\lambda)$ is dependence of the real part of the refractive index epidermis,
2. $\chi_j(\lambda)$ is dependence of the imaginary part of the refractive index epidermis

Study of the Optical Characteristics of Thin Layer of the Biological Sample for Case In vitro

Input parameters:

$n_j(\lambda) + i\chi_j(\lambda)$ is refractive index of the jth layer,
d_j is thickness jth layer,
M_1, M_2 are the radii of mirrors,
L is the mirror distance.
σ is the standard deviation of the profile of the rough boundary from the unperturbed boundary

Output parameters:

$\omega = \omega(d_j, n_j, i\chi_j, \sigma)$ are frequencies of the resonator eigenmodes

Experimental measurement:

$\chi_j(\lambda)$ is dependence of the imaginary part of the refractive index on simulated biological structures for jth layer.

Calculate:

1. $n_j(\lambda)$ is dependence of the real part of the refractive index epidermis
2. $\chi_j(\lambda)$ is dependence of the imaginary part of the refractive index epidermis

Simulation of the Thermal Processes for Case In vivo

Input parameters:

$n_j(\lambda) + i\chi_j(\lambda)$ is refractive index of the jth layer,
d_j is thickness jth layer,
m_1^k is refractive index of the cytoplasm
m_2^k is refractive index of the plasma membrane
a_1^k is the radius of the cytoplasm of kth particle,
a_2^k is the radius of the plasma membrane of kth particle,
c_0^j is specific heat capacity jth layer,
Λ^j is heat conductivity coefficient jth layer,
ρ^j is density jth layer,
μ is absorption coefficient,
E_o is radiant energy density,
τ_p is pulse duration,
ρ_{blood} is density blood,
T_{blood} is temperature blood,
$f(t, T)$ is the density of the flow of blood into the tissues,
is the reduced heat-transfer coefficient,
T_0 is the initial temperature,
ΔH is enthalpy of activation,
ΔS is entropy of activation,

Output parameters:

$T(\mathbf{r}, t, d_j, n_j, \chi_j, m_1^k, m_2^k, a_1^k, a_2^k, c_0^j, \Lambda^j, \rho^j, E_o, \tau_p, \rho_{blood}, T_{blood}, f(t, T), A, \Delta H, \Delta S)$
is spatial temperature distribution

Experimental measurement:

Threshold energy exposure.

Calculate:

1. Temperature distribution in the direction of the incident radiation
   (z axis) for a multilayer light-absorbing and scattering
   medium that simulates human skin and its components at various parameters
   value
   $(d_j, n_j, \chi_j, m_1^k, m_2^k, a_1^k, a_2^k, c_0^j, \Lambda^j, \rho^j, E_o, \tau_p, \rho_{blood}, T_{blood}, f(t, T), A),$

2. Estimation the kinetics of denaturation of tissue.

Index

© Springer International Publishing AG, part of Springer Nature 2018
K. Kulikov and T. Koshlan, *Laser Interaction with Heterogeneous
Biological Tissue*, Biological and Medical Physics, Biomedical Engineering,
https://doi.org/10.1007/978-3-319-94114-1

Printed in the United States
By Bookmasters